# IN THE MOUTH OF THE DRAGON

# IN THE MOUTH OF THE DRAGON

DEBORAH WALLACE, Ph.D.

AVERY PUBLISHING GROUP INC.
Garden City Park, New York

Cover Design: Rudy Shur and Martin Hochberg
In-house Editor: Karen Price

**Library of Congress Cataloging-in-Publication Data**

Wallace, Deborah.
In the mouth of the dragon: toxic fires in the age of plastics /
Deborah Wallace.
p. c.m.
Includes bibliographical references.
ISBN 0-89529-440-0
1. Plastics—Thermal properties. 2. Plastics in building—Fires and fire prevention—Case studies. 3. Combustion gases—Toxicology. 4. Smoke—Toxicology. I. Title.
TH9446.P55W35 1990
363.17'98—dc20 89-18239
CIP



Printed in the United States of America

10 9 8 7 6 5 4 3 2 1

# Contents

# Preface

As an ecologist with training in physiology and environmental and occupational health, since 1979, I have studied the health status of injured survivors and the autopsy reports of fatality victims of fires fueled either partially or totally by synthetic materials. Some of these fires were multifatal headline-grabbers, like the 1980 MGM Grand Hotel fire in Las Vegas and the 1977 Beverly Hills Supper Club fire in Covington, Kentucky, near Cincinnati. Other fires killed one person or two and, at most, got a paragraph in the local press.

Even before 1979, I needed to understand the nature of plastics and their behavior in controlled combustion because of my work for a corporation that intended to burn garbage to generate electricity. The plastics added worse pollutants than carbon monoxide to the stack effluent: acid gases, heavy metals, and toxic organic chemicals.

After approximately fifteen years of studying the public health impacts of burning plastics, the past ten being devoted to acute fume and smoke exposures, I knew that the hazards from plastics in fires should be brought to public attention. Because of the global atmospheric pollution that causes ozone layer thinning and the "greenhouse effect," the public is beginning to understand that the manufacture of plastics and other petrochemicals leads to apocalyptic environmental changes. Because of the trash disposal crisis, the public is beginning to understand that plastic is an intractable disposal problem. Most people, however, do not understand the environmental health risks of the middle stages of the life cycle of plastics: storage and use. One of the most frequent and serious of these risks is accidental fire and the generation of fumes and smoke from plastics.

The introduction and first two chapters of this book explain what plastics are; explore plastic's physical and chemical properties; examine the events of fires in buildings; and discuss the interaction between the nature of plastics, building design, and the dynamics of fires. Chapters Three through Eight each describe a relatively famous fire and the interactions that led to the deaths and injuries. Chapter Nine explains how we ended up with this widespread, inappropriate use of plastics, and Chapter Ten suggests actions that can roll back this tide. By the end of the book, the meaning of the book's title will be obvious: from the "fiery serpents" that beset the children of Israel in Deuteronomy to Fafner to St. George's opponent, the Dragon, by its nature, destroys life and property and is extremely difficult to overcome.

I have file cabinets full of autopsy reports and medical records. A person injured by fumes and smoke from heated plastic often is a patient for life and is hospitalized regularly. Some lessons must come out of all this death and suffering to prevent further losses. At this point in my profession, I have had to choose between writing this book to teach these lessons, and trying to forget them.

Many, many people made this book possible: my husband and son, who rooted for me during every case and during all the writing; my mentor at the Center for the Biology of Natural Systems (CBNS), Barry Commoner; my friends in the labor unions, including Lou Ackerman of Plumbers Local 299, Bob Gollnick who used to preside over the New York State Professional Firefighters, Tom Hunter with the California Plumbers, Alice Freund who had been with the Occupational Safety and Health Administration (OSHA) unit of District 37, George Browne of the New Jersey Firemen's Mutual Benevolent Association, and George MacDonald of the Transport Workers Union; conscientious building officials, like Bobby Brunner of Greenburgh and Fred Wilson of Albany; friends in the environmental movement, like Sister Margeen Hoffman, Pat Brown, Luella Kenny, Sister Joan Malone, and the gang from Ecumenical Task Force of Niagara; Bonnie MacLeod, who has been with the Sierra Club; Karim Ahmed of Natural Resources Defense Council; Pat Costner and Dave Rappaport of Greenpeace; Caroline Steiner, who has been with Environmental Action Coalition; my many attorney-clients, such as Stan Chesley, George LaMarca, Randy Duncan, Wendell Gauthier, Will Kemp, Dennis Maycher, Ivan Schneider, David Miller, Tom MacAlliley, to name a few; elected officials, like the folks at the Council of Insurance Legislators (COIL), New York State Assemblyman Richard Gottfried, former New York City Council President Carol Bellamy, Mayor David Dinkins, Dutchess County Executive Lucille Pattison, and United States Senator Edward Kennedy; and the many survivors who tried to teach me how it was to go through a plastics-fueled fire, and whose privacy will not be invaded by my naming them.

Deborah Wallace<br>
New York, New York

# Introduction

In the past, fire has been both a friend and an enemy to man. It has kept him warm, cooked his food, and lighted his way. It has also been a threat—man has become trapped in forest fires and fire has been used as a weapon. Fire was sometimes dangerous. Today, a new dimension of danger has been added. That danger is in the form of invisible gases that are released when man-made materials burn. With the growing number of highly flammable and toxic materials, man finds himself in a situation where he is surrounded by a potentially lethal environment.

Since the 1960s, this situation continues to grow unchecked, with very little public attention. In this book, it is our intention to document the dangers presented, why they are dangerous, and why they have gone unnoticed.

Today, we live in the mouth of the dragon. This dragon awaits us in every building, every bus and car, and every airplane. He is never far from us and slumbers uneasily, awakened by a kink in a wire, the fall of a lamp, or the careless flick of a match. The breath of the dragon could engulf any one of us, trapping us in an elevator or igniting at the start of a car engine or the passing of a rubbish fire in a vacant lot. We could be sleeping in a hospital room after an appendectomy, sitting in a box at an opera, or watching the World Series at home when the dragon attacks without warning.

This threatening dragon is plastics, and the dragon's breath, the smoke from plastics. What are those in charge of product safety doing to save us from this toxic smoke? What can we do to save ourselves?

Many consider plastic to be the perfect material. Plastics can be used in so many ways, and cost less than the real thing. You can buy vinyl concrete, vinyl siding for your house, vinyl window frames, polyvinyl chloride (PVC) plumbing pipe, acrylic windows, vinyl wallpaper, urethane foam-cushioned and acrylic-covered sofas and armchairs, nylon carpets, rigid polyurethane "wood" tables and bookcases, vinyl venetian blinds, and molded, rigid PVC chairs and tables. You can live in a house that is finished in plastic and furnished in plastic. Our gorgeous public buildings contain tons of plastic. Hotels, restaurants, government buildings, office towers, and even nursing homes are filled with it. And why not? It's usually cheaper than metal, glass, concrete, or traditional organic materials like wood, cotton, or wool. Why not use plastic?

The best reason not to use synthetics is that they pose risks to life and health that are qualitatively and quantitatively different from traditional organic and inorganic materials. In some cases, the dangers of plastics are much more severe, when they generate toxic or "exotic" fumes and smoke in fires and other situations of excessive heat.

Plastics are synthetic materials made from distillates of oil, coal, natural gas, or wood. These synthetic materials first invaded our homes and workplaces in large quantities in the 1960s. PVC wire insulation, nylon carpeting, and urethane foam mattresses and soft furniture were among the products first marketed widely in that decade. These plastic products were much cheaper than those made of the old materials and were marketed as "new," "innovative," and "modern." Even in the early days of mass marketing, several kinds of plastics were available.

No one thought to test these early synthetic polymers for their combustion toxicity. These products were virtually untested when they were put on the market. Instead, the public became the test animals.

In 1927, there was a disastrous fire in the Cleveland Clinic. Experts in the field of fire science and in the medical community agree that this was the first major fire fueled by synthetics. The smoke killed 135 people and injured many more. The fire grew rapidly, burned with unusual intensity, and generated massive volumes of dense, acrid, brown-black smoke. The fuel that made the fire so difficult to control and that was responsible for so many deaths was the nitrocellulose x-ray film used in the hospital. Nitrocellulose is a polymer.

In 1943, a fire at the Coconut Grove nightclub in Boston killed over 300 people and injured hundreds more. The National Fire Protection Association (NFPA) investigated this fire and, based on eyewitness accounts, the building construction, and the damage to the building, came to the following conclusions about the causes of injuries and deaths.

- Overcrowding had overwhelmed exit capacity.
- Blockage of paths to the exits by objects prevented escape.
- Combustible materials lined the walls and exitways.
- Marking and position of exits were inadequate.

The NFPA did not interview the emergency room staff and hospital physicians who cared for the injured survivors. Nor did that organization amend its report on the Coconut Grove fire after these physicians published several articles in *Annals of Surgery* that differed with its findings. The NFPA still takes the position that its report on the Coconut Grove fire is the conclusive truth.

The doctors at the receiving hospitals in Boston had no idea of the circumstances surrounding the fire. They did note that the most frequent type of injury was smoke inhalation. When they saw the soot-lined nostrils and mouths of the injured survivors, the doctors anticipated carbon monoxide poisoning as the overriding single medical problem. As the other symptoms began to arise, the doctors realized that the type of smoke inhalation injuries suffered by the survivors of this fire were different both in quality and in severity from the carbon monoxide-based smoke inhalation they were accustomed to treating.

Respiratory and cardiac failures resulted in delayed fatalities. Many of the injured developed delayed pulmonary symptoms such as shortness of breath, pneumonia, bronchitis, hyperproduction of mucus, and sensitivity to stress or exertion. From these unusual symptoms, and from the timing of the symptoms, the physicians concluded that a major factor in the deaths and injuries was the nature of the fuel. The decor of the nightclub included masses of nitrocellulose fake coconut fibers, which produced large quantities of nitrogen dioxide, a powerful respiratory irritant. To this day, few people outside the medical profession know of this fifth contributing factor in the Coconut Grove fire deaths and injuries.

During the 1960s and 1970s, the frequency of fires fueled by plastics increased greatly because of the increase in use of these materials, especially polyvinyl chloride (known as vinyl or PVC). Transportation fires, in particular, made headlines. Several airplane crashes occurred in which large numbers of passengers were killed by fumes and smoke, not by the trauma of the crash. Also noted was the tragedy aboard the USS *Enterprise* that killed so many sailors in New York Harbor when its electrical cables burned.

Official notice of the dangerous behavior of these new materials began to go public in the late 1960s–early 1970s. In 1969, the New York City Fire Department issued a safety bulletin that warned about the unusual risks associated with fighting fires in cable and wire insulation made of these new plastic materials.

In 1972, the report of the President's Commission on Fire in America (*America Burning: The Report of the President's Commission on Fire in America)* was published. For the first time, information that unequivocally questioned the safety of plastics in fires was made available to the general public. However, few people even knew the report existed. Included in the report is the following.

> What makes plastics relevant to our discussion of materials is not only that many of them have introduced hazards previously uncommon, but that they are sold and used without adequate attention to the special fire hazards they present. . . . This Commission recommends that the new Consumer Product Safety Commission give a high priority to the combustion hazards of materials in their end use. Specific needs are refined understanding of the destructive effects of smoke and toxic gases, development of standards to minimize these effects, development of labeling requirements for materials, and outright ban of materials in uses that present unreasonable risks.

In 1975, the NFPA tried to act on this issue by proposing the insertion of a requirement into the fire code that materials used in construction be no more toxic than wood. The Society of Plastics Industries (SPI) successfully sued to stop this insertion into the fire code.

The Society of Plastics Industries and individual manufacturers of urethane foam reached a settlement agreement with the Federal Trade Commission in 1974–1975. The Federal Trade Commission (FTC) had found misleading advertising that claimed that urethane foam was fireproof and self extinguishing. The commission found that the deaths of two children in a fire at a Jewish community center were caused by the flammability and fire toxicity of the urethane foam in cushions at the center. The center, along with the children's parents, sued the manufacturers and the SPI for misleading sales statements and for selling a dangerous and defective product.

Unfortunately, the FTC settlement agreement did not regulate the use of the product. It merely required that the urethane foam industry place labels on the product, educate its salespersons about the fire hazards of the product, and submit an annual report on the program. Thus, the industry formed a task force to organize these activities and "rehabilitate" the product in the public mind. This urethane foam safety group of SPI became the model for other similar groups for other dangerous plastics. The settlement agreement, far from protecting the public, became the trigger for organizational activity that enabled the industry to market unsafe products and to expand markets into ever less appropriate uses. Chapter 9 contains more about this issue.

The utter weakness of the FTC settlement led to further deaths from the unlimited use of urethane foam. In 1983, twenty-nine people died in a prison in Biloxi, Mississippi from the toxic smoke produced by urethane foam padding that caught fire. Unfortunately, no progress had been made—and no lesson had been learned—since the 1977 Columbia County, Tennessee prison urethane-fueled fire that killed forty-three people.

Indeed, the 1977 Tennessee prison fire was one of a new series of plastics fires that killed or injured large numbers of people. These fires include the 1975 New York Telephone fire that injured 239 out of 700 firefighters who battled the blaze (PVC fueled); the Beverly Hills Supper Club fire of 1977, in which 165 were killed (a PVC and electrical fire); the 1978 Cambridge, Ohio Holiday Inn fire (10 died from smoke from PVC and nylon); the 1978 Younkers Brothers Department Store fire (10 died in this PVC-electrical fire); the 1980 MGM Grand Hotel fire (85 died; the fire was largely fueled by plastics); the 1980 Stouffer's Inn fire (26 died, and the fire was primarily fueled by PVC and nylon/wool); the 1983 Westgate Hilton fire (12 died from smoke that came mainly from PVC and urethane foam); and the 1983 Fort Worth Ramada Inn fire (5 died from PVC and nylon fumes). This book will examine a few of these fires in detail to demonstrate how the particular flaws in the synthetics present in the buildings contributed to deaths, injuries, and damage.

The fires listed above occurred in the United States. Other countries also struggle with the lethal hazards of synthetic materials used in building furnishings and finishings. In 1967, approximately 200 people died in an electrical fire in a Brussels department store. Hotel fires in Tokyo, Manila, Moscow, and São Paulo injured and killed scores of people. The 1983 Dublin Stardust Lounge fire killed 125, and a 1984 fire in a Turin, Italy movie theater killed 64.

The large multi-fatal fires in public places aren't the only ones in which people are killed by fumes and smoke from plastics. Most homes now contain significant quantities and many types of plastics. Mattresses and sofas involved in falling-asleep-while-smoking fires usually contain urethane foam. Victims of these fires often show elevated levels of cyanide in their blood, a result of the smoke and fumes emitted by urethane.

Polyvinyl chloride (PVC) probably constitutes the single most abundant synthetic material in the home. Wallcovering, electrical insulation, wastebaskets, garbage cans, molded furniture, drain pipes, and vinyl siding are among the household items made from it. PVC also decomposes more readily than most other common plastics.

The following is an account of a fire that killed two children. It comes from a report written for the plaintiffs in the court suit against the manufacturer of the PVC and wood wallcovering.

> On February 4, 1982, a fire occurred in Warren, Michigan which resulted in the deaths of two children and injuries to the three survivors. An investigation of this fire was conducted by reviewing the house itself, photos of the fire damage, and the statements of witnesses including the adult survivor, and the fire marshal. Samples from the paneling noted by the fire marshal as the path of the flame were analyzed by infrared scan of the surface film which was revealed thereby to be poly-

> vinyl chloride. This film was further tested by thermal decomposition and shown to release HCl (hydrogen chloride) in large volumes. The panel burned very readily also.

The plastics industry and the governmental agencies responsible for regulation of these products know the details of how these materials behave in fires. Both the mass deaths and the individual residential fire deaths that result from this behavior can be prevented.

This book will describe the toxic hazards of plastics in terms that the general public can understand, as opposed to the confusing terms and wording practiced by industry and governmental scientists. It will also provide vitally needed public education about the health risks that are posed by plastics in fires and other situations of excessive heat. The fact that such public education is still needed should shame governmental agencies like the Consumer Products Safety Commission and those quasi-governmental agencies that receive public funding, like the National Fire Protection Association. Considerations of public health and safety have languished in this matter.

# CHAPTER 1

# Elements of Death

Have you ever noticed that polyester clothing pills; that vinyl shower curtains crack; and that polyethylene bags rip? There is a relationship between this pilling, cracking, and ripping and the deaths in the fires described in later chapters of this book. Plastics, by their nature, cannot actually replace the materials they are often substituted for, because they begin to decompose even as they are manufactured. The way plastics are made and the elements from which they are made explain why they are predisposed to decomposition.

## THE BUILDING BLOCKS OF PLASTICS

The building blocks of plastics and synthetic fibers come from natural gas, coal, and wood, but the major source is oil. The raw natural gas, coal, wood, and oil are distilled. This process of distillation results in intermediate products that are again modified before being polymerized or packaged for marketing as a liquid. Common products made from coal derivatives include asphalt, tar, creosote, and cresol. Products made from wood derivatives include rayon, turpentine, pine tars and oils, and wood alcohol.

The raw fuels natural gas and oil are distilled and "cracked" to make synthetics. (In the petroleum industry, the term "cracked" means breaking large molecules into smaller ones.) These raw fuels contain mixes of organic molecules that are of different sizes and structures. These molecules are "sorted" by the distillation process. When a product is distilled, it is boiled, and lighter, smaller molecules are separated from heavier and larger ones. The first thing that boils off has the lower boiling point and is comprised of lighter and smaller molecules. The heavier and larger molecules are then further cracked to yield more smaller, lighter molecules.

The result of all this distillation and cracking is organic chemicals, which are chemicals that are made of carbon and other elements, typically hydrogen, oxygen, and nitrogen. They become the building blocks for the "new" synthetics such as plastics, glues, and pesticides, and for their helping additives, such as fire retardants and plasticizers. These building blocks are often modified before being formed into the product. Chlorine, phosphorus, nitrogen, or another element may be added to change the properties of one of these substances and make a new molecule. For example, chlorine is incorporated in ethene to produce a monomer called vinyl chloride monomer, the building block of a polymer called polyvinyl chloride or PVC.

## POLYMERIZATION

After the basic molecular building blocks are manufactured, they are turned into plastic resin. This process is called **polymerization** and results in the creation of a **polymer.** A polymer is a large, organic, chain-like molecule made of repeating units of smaller molecules. "Poly" means many and "mer" means units: many units. Polymerization generally involves heating the building blocks in a vessel with "helper" chemicals called catalysts until the building blocks join to form long chains. Even with the catalysts, a great deal of heat is used in the polymerization process. Because of this heat, the long chains, even during manufacture, may decompose slightly and have defect points along them. The defect points are in the chemical bonds, which absorb the energy used in the manufacture process. The law of conservation of energy states that the amount of energy in a system *after* the reaction is the same as the amount of energy *before* the reaction. The large amounts of energy (heat), thus, must go somewhere. They go into the bonds between the atoms of the plastics and are stored there. But, as stated before, Nature does not favor this gain of energy—Nature favors low energy chemical bonds, and high energy bonds tend to release their energy by breaking. These are defect points. When something is manufactured in a high heat environment, this itself will break some of these unstable bonds. Although polymer scientists and engineers have striven to reduce the number of defect points, they have not been able to completely eliminate them from synthetic polymers.

The physical and chemical defects that are produced by ordinary processes in the manufacture and use of plastics demonstrate the fragile and unstable character of these long chains of molecules that are joined by high energy chemical bonds. When the resin is further processed to become the finished marketable product, additional defect points are created because the product is again heated and handled.

Plastics aren't true solids. Chemical physicists place them in the category of "viscoelastic fluids," and they are closer in behavior to yogurt than to a true solid such as iron. Given enough time, most plastics will flow or "creep." **Creep** is a change in the physical dimensions of a piece of plastic, caused by the movement of the molecules past each other. Usually, the plastic becomes thinner, longer, and wider—it stretches. An example of creep can be seen in Figure 1.1. Eventually, creep causes physical weakening, and leads to easy decomposition. Physical forces such as pressure, torsion, or shear will speed this flow, as will increases in temperature.

**Figure 1.1.** An example of creep. This is PVC cable.

Physical forces, exposure to extreme cold, or contact with certain chemicals will cause plastics to crack. This is called **environmental stress cracking.** Sunlight, changes in temperature, and air pollutants render some plastics brittle and/or oxidize them so that they lose strength and resilience, and their appearance changes. Many of the environmental conditions that lead to the physical failure of plastics also predispose them to unraveling chemically.

## SLOW DECOMPOSITION

During normal use, many synthetic materials decompose slowly, beginning at the defect points created during the manufacture of the resin and the processing of the finished product. Polymer scientists call this slow decomposition "aging." This aging process potentially affects the health of all the millions of people who live or work in new or rehabilitated buildings. These buildings are tightly insulated and recirculate their air, with little dilution from the outside. The design of the buildings is supposed to promote energy conservation. The synthetics in the buildings volatize or decompose, and fumes accumulate in the recirculating air, eventually causing what has come to be known as "office worker illness." Some of the symptoms of this illness are susceptibility to respiratory disease, headaches, sore nose and/or throat, irritated eyes, fatigue, skin irritations, and personality changes.

Besides fire and new or renovated tightly closed buildings that contain much plastic, the public is exposed to fumes and particles from decomposing plastics in other circumstances also. Cars left in the sun on a hot day may reach temperatures over 125°F (52°C) inside the car. The normal rate of slow decomposition is accelerated by the combination of heat and sunlight, and the acceleration results in that characteristic odor you smell when you open the car door. Cheap lamps often are made with plastic shades that slowly char where the shade is nearest the bulb. This also produces a characteristic odor. Urethane foam mattresses and sofa cushions actually powder and emit a fine dust after several years of use. The odors and the dust contain the chemical products of decomposition. The contribution of these plastics-related exposures to incidence of disease is unknown, although the urethane powder has been shown to trigger attacks in asthmatics. The exposures are now so frequent, however, for members of the general public, that a cumulative effect must be considered possible.

During slow decomposition, which with many synthetics occurs continuously, the plastic will be in contact with the gases it emits. On some plastics, such as PVC, these gases act to facilitate further emission of gases. Such facilitation by products of decomposition is called **autocatalysis**—the process of decomposition in which gases from the material aid in the decomposition of that material. Three factors cause physiologically significant quantities of gases to arise from plastics:

1. The pre-existence of chemical decomposition even before unusual heating.
2. The intertwining of chemical and physical degradation. Chemical degradation means loss of parts of the molecule. Physical degradation is a change in physical characteristics such as strength, dimensions, color, integrity (cracks or holes), or electrical resistance, etc.
3. Autocatalysis.

Because of these factors, gases may emanate from the plastic even before the temperature of quantitative (rapid and predictable) decomposition is reached in the heat buildup.

The fundamental skeleton of plastics generally consists of carbon or carbon and oxygen. The products of early decomposition are those parts that extend from the main carbon chain. These are generally acidic groups, basic groups, or an organic group which can become acids or aldehydes in the air. These chemicals break off from the main molecule. Thus, the products are not the inert carbon, but the acidic or basic parts. The deadly acid gas hydrogen chloride (HCl), for example (in aqueous form, it is called hydrochloric acid), comes off PVC so quickly and so easily that polymer scientists call it "unzipping."

One of the normal aging processes of synthetic polymers is chemical decomposition. As stated before, even at room temperature, many plastics emit their additives (fire retardants, stabilizers, colorants, lubricants, plasticizers) and easily broken-off parts of the polymer itself into the air. Heat accelerates the decomposition which, in turn, greatly accelerates the degradation of physical properties. The early stages of a fire, thus, include interrelated chemical and physical degradation. There are few visible signs of change unless one knows how and where to look for them.

Many of us consider flames the signal that a fire has begun. Combustion scientists think of a fire as beginning long before actual flame is present. Flame results from a process that begins with a relatively low level of heat. With plastics, the first stage of a fire is invisible; heat causes the molecules to slip and slide, and causes a great acceleration of the aging processes of oxidation, flowing, and loss of additives. Eventually, the heat builds to the softening point and then to a melting point. In plastics, physical changes such as melting, deformation, expansion, and loss of strength are part of the first stage of a fire. Changes in electrical properties may also appear.

## QUANTITATIVE DECOMPOSITION

As the temperature surrounding the plastic rises, the decomposition accelerates. The temperature eventually reaches a special level called the **temperature of quantitative decomposition** during which large and predictable quantities of gases are emitted by the polymer. Both natural and synthetic polymers each have a characteristic temperature at which quantitative decomposition begins.

This weight loss (emission of large, predictable quantities of gases) occurs only at or above these characteristic temperatures. The 480°F (250°C) oven temperature used to roast a chicken on a vertical rack will quantitatively decompose PVC and release clouds of hydrogen chloride gas. Yet PVC doesn't actually burn until it reaches about 1112°F (600°C). Fire-retarded urethane foam begins to decompose at only the temperature of boiling water (212°F, 100°C), quantitatively decomposes at only about 660°F (350°C), and burns at

800°F (427°C). There is a long period of decomposition during which acid gases and/or asphyxiants such as cyanide leave the plastic in concentrated clouds. During this period, people can become disoriented or disabled by fumes and smoke and will not be able to escape by the time the flames actually begin.

Most plastics have a number of chemicals added to them, such as fire retardants, stabilizers, lubricants, plasticizers, and colorants. These additives can only modify the problems somewhat. Fire retardants cannot alter the *decomposition* temperature—they can only delay the outbreak of *flames.* With time, many of these additives simply evaporate or leach out of the plastic. This results in deterioration not only in the performance of the product, but in its chemical stability as well.

Once quantitative decomposition begins, gases are emitted from the plastic with amazing rapidity. Once the decomposition temperature is reached, lethal concentrations of hydrogen chloride appear within two to three minutes at a distance of one to two feet from a four-ounce piece of PVC. These lethal gas concentrations are invisible, although in another two minutes or so, a steam-like, innocent-looking haze rises from the heated PVC.

Many other types of plastic release a thin haze during quantitative decomposition. This harmless-looking fog is usually highly concentrated and hazardous. Depending on the plastic, it may be acid gas, irritating aldehydes, or cyanide. The plastic becomes a true toxic hazard during quantitative decomposition when there is no flame to warn anyone who may be nearby.

Unlike traditional materials, the quantitative decomposition stage for plastics usually involves very rapid release of large quantities of toxicants. This is because plastics are high energy and unstable polymers compared with natural polymers such as wood, cotton, wool, and hair, which must be synthesized by organisms on an energy budget and are therefore low energy and stable. Nature prefers low energy chemical bonds; high energy bonds tend to release their energy by breaking and are therefore unstable. Also unlike traditional materials, the temperature of quantitative decomposition for many plastics is less than half their ignition temperatures. Because of this large temperature difference, there is a long period of time when gas is emitted without the warning presence of flame.

Generally, the gases emitted during the decomposition stage of a fire are more toxic than those emitted during actual burning. Thus, in many fires, the decomposition stage is the real killer. It is a killer because of its insidious and invisible nature, its high toxicity, and the long period of time between attainment of quantitative decomposition temperature and ignition temperature. In this respect, fire-retarded plastics are worse than non-fire-retarded plastics.

## HIGH TEMPERATURE OXIDATIVE PYROLYSIS

Another preflame stage has been identified by combustion scientists. This stage occurs between the beginning of quantitative decomposition and actual burning and is called **high temperature oxidative pyrolysis.** (The word pyrolysis refers to chemical change caused by heat.) During this preflame stage, physiologically significant quantities (amounts that can adversely affect health) of carbon monoxide arise from the plastic. Once the bonds of the main carbon skeleton break from the intense heat, large quantities of soot are emitted. The toxic hazards of the high temperature pyrolysis stage are different from the toxic hazards that arise from the earlier lower temperature initiation of quantitative decomposition. In the high temperature pyrolysis stage, carbon monoxide, irritating and intoxicating organic chemicals, and large quantities of soot are present.

In most laboratory and real fires fueled by plastics, the chemicals emitted during quantitative decomposition linger in the air or continue to be emitted during high temperature pyrolysis, and during the pyrolysis stage soot is also formed. These emissions mix together and the mixing of the acid gases emitted during quantitative decomposition with the dense soot emitted during high temperature pyrolysis poses a particularly treacherous toxic hazard: the acid condenses onto the soot and exposes people to both gas and particles. The gas is inhaled and generally attacks the upper respiratory tract. But because it is solid, the acidic soot becomes deposited in the lower respiratory tract, in the air sacs and the bronchioles of the deep lung. Here, the acid reacts with and destroys the tissue necessary for the fundamental process of respiration, that is, the taking of oxygen into the blood and the release of carbon dioxide from the blood.

During high temperature oxidative pyrolysis, carbon monoxide is produced and the respiratory tract is exposed to corrosive acid in the forms of both gas and particles of soot. Carbon monoxide acts on hemoglobin (the oxygen-carrying protein in red blood cells), binding it to form **carboxyhemoglobin.** Carbon monoxide (CO) has a much greater affinity for hemoglobin than does oxygen. The carbon monoxide in the bloodstream attaches to the hemoglobin, therefore blocking oxygen from being bound to the hemoglobin and transported to the tissues. Tissues such as the brain, heart, and kidneys, which require high levels of oxygen, are especially subject to damage from both the elevation of the carboxyhemoglobin level in the blood and the acid-producing lung damage. Either carbon monoxide or acid damage alone can cause death by asphyxiation.

Acidic gas and particles injure tissues in three ways: direct chemical burns, accumulation of fluids into the tissue spaces from blood vessels and capillaries (edema), and hemorrhaging. Lung edema and hemorrhaging essentially drown a person in his own body fluids and no exchange of respiratory gases (oxygen and carbon dioxide) is possible. When the lung tissue is attacked by the acid, the "cement" holding the two layers of air sac cells together, called the basement membrane, is partially dissolved. These two layers then separate and cannot function. Then, the protein from the basement membrane mixes with the serum proteins from the edema and hemorrhaging and forms a hyaline membrane, which becomes a formidable barrier to gas exchange. If the concentration of acid gas released by the plastic is high enough, the damage to deep lung tissue can occur during the very early stages of quantitative decomposition.

## COMBUSTION

By the time the temperature has reached ignition level, anyone present often has inhaled physiologically significant amounts of acid gases, carbon monoxide, and soot. The pattern of injuries and deaths in fires is largely determined by the timing of the fire stages, with their characteristic releases of chemicals and soot particles. About 80 percent of all fire deaths come from inhalation of fumes and smoke, not from burns. Although the dead are often burned, the flames reach the bodies after death.

Actually, fire brings relief from some of the toxic hazards: much more of the material is being emitted as carbon dioxide during flaming than during pyrolysis, and less as carbon monoxide and larger organic chemicals. The main danger from carbon dioxide is that high concentrations exclude oxygen and can cause smothering. In small doses, carbon dioxide is not very toxic. Indeed, our own respiration produces it and we are exposed to it in our tissues constantly. However, large concentrations of carbon dioxide can deprive an atmosphere of oxygen and result in death.

## THE BEHAVIOR OF CERTAIN PLASTICS IN FIRES

Given the understanding of the fire stages that blend into each other and that are described above (1. Loss of strength and slow decomposition—this stage is not exclusively a pre-fire stage; 2. quantitative decomposition; 3. high temperature oxidative pyrolysis; 4. flame and true combustion), we shall apply these stages for some very common plastics. These are plastics that can create unusual environments, both qualitatively and with respect to high concentrations of nontraditional decomposition and combustion products. It is important to remember that the plastics we shall discuss are very common.

## Polyvinyl Chloride

For many years, the dominant plastic on the market has been polyvinyl chloride (PVC or vinyl). PVC is found in electrical insulation, plumbing pipe, plastic-film wallcovering, molded furniture, most "leather-look" upholstery, house siding, and bleach bottles, among other things. Under normal use, PVC will slowly decompose and give off hydrogen chloride, especially if it is exposed to ultraviolet light, temperature fluctuations, or physical forces. PVC that is a few years old has many defect points that predispose it to further and easier decomposition. If PVC is heated to only 212°F (100°C), the boiling point of water, it will lose some weight, but not enough to cause physical harm. This weight is almost entirely hydrogen chloride. Heating PVC to 480°F (250°C) will trigger quantitative decomposition—the emission of massive clouds of hydrogen chloride and trace amounts of vinyl chloride monomer that had not been incorporated into the long carbon chain. Potentially 58 percent of the weight lost by rigid PVC during the various fire stages can be attributed to hydrogen chloride. For a four-ounce, completely burned piece of PVC conduit or pipe, over two ounces will evolve into the air as hydrogen chloride. The concentrations of acid gas in the air near a piece of decomposing PVC will rise to a level that will cause serious injury or death very quickly, within two or three minutes.

Besides the acidic gas hydrogen chloride, a wide variety of chlorinated and nonchlorinated organic chemicals evolve from PVC during high temperature pyrolysis and combustion: benzene, toluene, formaldehyde, chloroform, chlorinated biphenyls, dioxins and dibenzofurans, and many others. In all, three hundred chemicals have been detected. Although at most 10 percent of the rigid plastic weight loss is organic chemicals, when hundreds of pounds are heated, the quantity of these organics may be great enough to have a negative effect on health.

Firefighters and the elected officers of their labor organizations have begun to question the health effects of hydrogen chloride, carbon monoxide, and the organic chemicals mentioned in the preceding paragraph. The emission during fires of benzene, chlorinated dioxins, and dibenzofurans, known carcinogens, appears to explain the high frequencies of leukemia, laryngeal and colon cancer, and of the rare soft tissue cancers found in many firefighters at relatively young ages.

Plasticized PVC emits high concentrations of the plasticizer, phthalates, as well as everything the rigid form, which contains no plasticizer, emits. Phthalates are irritants and cardiotoxins (heart poisons) and are very combustible, posing a threat of explosion or secondary fire at locations remote from the primary fire. The gas clouds may travel, touch a hot surface, and burst into flames.

**Urethane Foam**

Another commonly used plastic is urethane foam, which has a variety of formulations that are both fire retarded and non-fire-retarded. Although the non-fire-retarded forms have a very poor ignition and flame-spread performance, the fire-retarded forms decompose and burn at rather ordinary temperatures as well: most forms begin decomposing at temperatures as low as that of boiling water (212°F, 100°C), and enter quantitative decomposition at 570°–660°F (300°–360°C). Urethane foam emits the colorless and poisonous gas hydrogen cyanide during decomposition and combustion. As in the case of many plastics, urethane foam emits organics, carbon monoxide, and soot during the high temperature pyrolysis stage, which occurs before ignition. Benzene, acetaldehyde, and methanol are among the organics that are emitted by urethane foam. Some formulations also emit very strong irritants such as ammonia or toluidine di-isocyanate. The decomposition pattern of urethane foam is slightly different from PVC: with urethane foam, the higher the temperature, the faster the release of hydrogen cyanide, even during combustion. With PVC, by the time combustion begins, the peak of hydrogen chloride release is usually past.

Urethane foam has several uses, such as mattresses and upholstered furniture cushioning, thermal insulation on buildings, and car seat cushions. It is usually what is used in the cheapest mattresses and upholstered furniture. The inexpensive chairs and couches that can unfold to become beds contain urethane foam. Because it is so inexpensive, urethane foam mattresses are often found in prisons, dormitories, hotels, and nursing homes.

Fires in urethane foam may result in deaths by asphyxiation from both hydrogen cyanide and carbon monoxide. Besides combining with hemoglobin much the way carbon monoxide does, cyanide combines with enzymes called cytochromes, which are necessary for respiration on the cellular level. Hydrogen cyanide, ammonia, and toluidine di-isocyanate are irritants that can damage the tissue of the respiratory tract. (**Irritants** are chemicals that cause a sensation of burning, reddening, and inflammation. **Corrosive irritants** kill tissue and cause severe inflammatory reactions.) In a urethane foam fire, respiration is attacked at three levels: at the tissues where gas exchange occurs, at the hemoglobin in the blood, and at the cytochromes in the body cells.

## Polystyrene

Yet another common plastic is polystyrene. Polystyrene is used as hot/cold disposable drinking cups, packing material, cushioning material inside electric appliances, building insulation, and the trays on which large cuts of meat are wrapped for sale in supermarkets. Polystyrene, like PVC and urethane, begins to decompose at 212°F (100°C). It reaches quantitative decomposition at 390°–578°F (200°–300°C), depending on formulation. It burns at about 840°F (450°C), also depending on formulation. Polystyrene is different from

PVC and urethane because as it is manufactured, no potentially troublesome elements like the chlorine of PVC and the nitrogen of urethane are added. Theoretically, it should cause very little trouble. However, in lab toxicity tests, expanded polystyrene (plastic foam) seemed slightly more toxic than urethane or PVC formulations.

Compared with PVC and urethane, burned polystyrene emits high concentrations of aromatic hydrocarbons: styrene, modified styrene, benzene, ethyl benzene, and phenols and phenyls. These are central nervous system depressants as well as irritants. They can kill in the same manner that narcotics kill. Several combustion toxicologists exposed laboratory animals to the fumes of burning polystyrene and performed autopsies on the dead lab animals. The researchers were amazed to find no evidence of any tissue damage other than slight respiratory irritation, not enough to explain the deaths. The central nervous system may simply turn off vital functions in these deaths.

The fire behavior of several widely used plastics, including PVC, urethane, and polystyrene, is summarized in Table 1.1. Note that most of these products emit benzene during high temperature pyrolysis and combustion. Benzene presents a double danger—it is very combustible and can spread the fire rapidly, and it causes cancer. Benzene emitted during fires is suspected of causing the high rate of leukemia among firefighters, but the hazard it poses to the ordinary citizen has received very little attention. The ever-widening use of plastics may mean that many people are being repeatedly exposed to plastics-derived benzene.

## HAZARDOUS WASTE BY-PRODUCTS OF PLASTICS

Pesticides and plastics have common ingredients and common hazardous waste by-products. The famous Love Canal and Hyde Park toxic dumps (both near Niagara Falls, New York) from Hooker Chemical and Plastics Company came from one-site manufacturing of several chlorinated products. Among these products are DDT (pesticide), Mirex (pesticide),

**Table 1.1.** Some Common Plastics and Their Fire Behavior.

| Plastic | Commonly Found Products | Beginning Temperature of Quantitative Decomposition |
|---|---|---|
| **Polyvinyl chloride (PVC)** | *Plasticized:* Wire insulation, wall covering, leather-look upholstery | 464–536°F (240–280°C) |
| | *Rigid:* Electrical conduit, plumbing pipe, molded furniture | 480°F (250°C) |
| **Urethane Foam** | *Flexible:* Pillows, mattresses, soft furniture stuffing | 212–390°F (100–200°C) |
| | *Rigid:* Building insulation, some wood-looking furniture | 300–390°F (150–200°C) |
| **Acrylonitrile-butadiene-styrene (ABS)** | Plumbing pipe, upholstery, computer casing, luggage | 570°F (300°C) |
| **Polystyrene** | Hot/cold drinking cups, fast-food packaging, other food packaging, insulation | 390–570°F (200–300°C) |

| **Emissions During Quantitative Decomposition** | **Beginning Temperature During Ignition** | **Emissions During Ignition*** |
|---|---|---|
| Hydrogen chloride[†], phthalate[§] | 1,110°F (600°C) | Carbon monoxide[‡], benzene[§], organics[§] |
| Hydrogen chloride[†] | 1,110°F (600° C) | Carbon monoxide[‡], benzene[§], organics[§] |
| Large quantities of organics[§] (50% weight of foam), hydrogen cyanide[†] | 750–840°F (400–450°C) | Nitrogen oxides[†], benzene[§], carbon monoxide[‡], organics[§] |
| Organics[§], hydrogen cyanide[†] | 840° F (450° C) | Nitrogen oxides[†], benzene[§], carbon monoxide[‡], organics[§] |
| Hydrogen cyanide[†‡], *organics*[§] | 930° F (500° C) | Nitrogen oxides[†], carbon monoxide[‡], benzene and other organics[§] |
| Styrene[§], phenyls[§], diphenyls[§] | 840° F (450° C) | Carbon monoxide[‡], phenols[§], other organics[§] |

**Table 1.1.**—*Continued*

| Plastic | Commonly Found Products | Beginning Temperature of Quantitative Decomposition |
|---|---|---|
| **Teflon** | *Plasticized:* Wire insulation fire alarm on telephone cable | 570°F (300°C) |
| | *Rigid:* Credit cards, teflon cookware, protective liners (such as on the Statue of Liberty) | 930–1,110°F (500–600°C) |
| **Nylon** | Carpets, upholstery, clothing, gears | 660–750°F (350–400°C) |

*Emissions in addition to those begun during quantitative decomposition.

†These chemicals are corrosive irritants and cause the following injuries: direct respiratory tissue death from corrosion, indirect respiratory tissue death from the inflammatory reaction, lung edema and hemorrhage, chemical pneumonia and bronchitis, susceptibility to respiratory infections, permanent abnormal lung functions, skin scarring and sensitization, eye damage, neurological and vascular reactions.

lindane (pesticide), PVC (plastic), and PCBs (plasticizer, fire-retardant, and insulator). These products were made at the one manufacturing site because of many common feedstocks that are necessary for all these products. (Feedstocks are the chemicals needed for the manufacture of these products.) Before 1974, PCBs (polychlorinated biphenyls) were used as a plasticizer-fire retardant as well as an electrical insulation/coolant. By-products from all of these manufacturing operations included dioxins, dibenzofurans, chlorinated benzenes, and other organic chemicals.

| Emissions During Quantitative Decomposition | Beginning Temperature During Ignition | Emissions During Ignition* |
|---|---|---|
| Hydrogen fluoride[†], phthalate[§], fluorinated organics[§] | 1,110–1,290° F (600–700° C) | Carbon monoxide[‡], benzene[§], other organics[§] |
| Hydrogen fluoride[†], fluorinated organics[§] | 1,290–1,470° F (700–800° C) | Carbon monoxide[‡], benzene[§], other organics[§] |
| Hydrogen cyanide[†‡], ammonia[†] | 930° F (500° C) | Carbon monoxide[‡], nitrogen oxides[‡], organics[§] |

[‡]These chemicals are asphyxiants. Survivors may have damage in organs needing high levels of oxygen: heart, brain, kidney, and liver.

[§]Organic chemicals usually affect the nervous system. Phthalates are also heart poisons. Benzene causes blood cell abnormalities, including leukemia. Many organics poison the liver and cause cancer. Chlorinated organics also may cause reproductive system disorders. Fluorinated organics cause tissue corrosion and progressive ulceration.

Pesticides and PCBs in a fire also pose significant hazards because of the chemical decomposition that occurs to make them more toxic. For example, when PCBs are heated, especially in air, they form chlorinated dibenzodioxins and dibenzofurans that are much more toxic than the original PCB. PCBs, in a manner similar to polyvinyl chloride, will also break down into hydrogen chloride and organics. The most famous and most studied PCB fire is the 1980 Binghamton, New York State office building fire. Many of those involved in the fire (office workers, firefighters, police officers) were contaminated with PCBs and dioxin. Now, cancer rates are higher among this group than among others. The firefighters and police officers who were at the scene now have high dioxin levels in their fat tissues. Many transformer explosions and fires have exposed thousands of workers around the country to these highly toxic chemicals. These exposures may be as serious as that of the emergency workers in the Binghamton event.

Other pesticides may contain nitrogen (for example, pyrethrin and carbaryl) or phosphate (malathion and parathion). Like plastics containing nitrogen or phosphate, these synthetics usually become much more toxic under high-heat conditions. Although some form of cyanide is likely to dominate the smoke toxicity of nitrogen-containing pesticides, other very toxic species will also be present such as ammonia, nitrogen dioxide, and nitrosamines. Nitrosamines have been shown to cause cancer in laboratory animals. Like plastics that contain phosphate, phosphorylated pesticides may thermally react to form nerve-gas-like heat products.

Safe storage of these pesticides in large quantities should be a high priority of federal and state agencies. However, the United States Department of Transportation, the Environmental Protection Agency (EPA), and Occupational Safety and Health Administration (OSHA) have softened their standards and allowed many toxic chemicals to be transported, stored, and disposed of in plastic containers. Many of these containers are made of polyethylene, which is almost as flammable as candle paraffin (to which it is chemically related). Again, these decisions were rendered under intense lobbying and threats by the plastics industry and its trade associations such as the Plastic Drum Institute of the Society of Plastics Industries. One federal agency that has retained some sanity is the Bureau of Mines, which tested plastic pails and found them too flammable to be allowed in mines as containers of combustible materials.

Because of the long struggle of Rachel Carson, the author of *Silent Spring,* and of many wildlife and environmental health research scientists, most of the public realizes at least vaguely that pesticides pose some risks to health. Their purpose is to kill insects, rodents, and other pests. If they can kill these other life-forms, we have little trouble understanding that they may cause problems for humans also, since we are just another life-form. It is now standard to use caution when handling and administering pesticides. We are no longer surprised when town meetings are held to discuss the dangers of tree-spraying, and take these considerations and deliberations very seriously. The health hazards of plastics, however, still remain unknown to most people. The large fires involving burning plastics have raised some concerns about the issue, but many are confused about the toxicity of the materials. This confusion has led to an inability and unwillingness to disentangle the obvious lies, the obvious truths, and semi-truths. Most people don't even know that there is a close relationship between plastics and pesticides, or that they are often manufactured on the same site with the same feedstocks. In a way, this ignorance and confusion makes plastics more dangerous than pesticides, because the same cautious approach simply does not exist for plastics as for pesticides. This is tragic, and should be changed.

## THE ROLE OF INDUSTRY AND GOVERNMENT

Those with authority in industry and government have known for decades that there are inherent problems in the normal use of plastics as consumer goods. In 1951, the National Bureau of Standards held a symposium entitled "Degradation of Polymers" during which polymer scientists from industry and government expressed their concerns about the instabilities of many common plastics and the poor product performance resulting from these instabilities. Seminars were held in the 1960s and 1970s on the same theme. At a 1978 conference sponsored by the American Chemical Society, the PVC chemists made nearly the same complaint that was made in 1951 about the easy decomposition of plastics.

After more than twenty-five years, the problem remained unsolved. But aggressive marketing had expanded the use of plastics, often into inappropriate areas.

Although industry and government beat the drum and remained gung-ho for the use of plastics, the consumer of today faces the same basic problems as in decades past: if you drop a plastic yogurt container, it will crack, but if you drop one of waxed cardboard, it will merely dent; an old polyester dress will become unwearable, but a wool or cotton one never will; vinyl shower curtains become brittle and crack; nylon carpets stretch and bunch up; and PVC cable sheathing can't protect electric wiring from a nail hammered into a wall. Many common plastics cannot truly replace the materials they are being sold to replace!

The impact of product failure is primarily economic, but it also costs lives. For example, two New York City firefighters died in June, 1980 because they had to swing off of a burning building using a nylon rope. The rope stretched, and its cross-sectional strength was reduced, so that the edge of the roof actually cut the rope. A manila rope would not have stretched as badly or lost as much cross-sectional strength—this behavior is only inherent in nylon. The nylon rope complied with the Rope Association's standards, but the Fire Department hadn't taken notice that under intensive lobbying by the plastics industry, the Rope Association had weakened its standards. The result of this lobbying by the plastics industry was the manufacture and use of nylon rope that would have failed to pass the original standards. The plastics industry takes our money and our lives by lobbying, by intimidation, and by massive spending to water down the standards and regulations that are meant to protect us.

*CHAPTER 2*

# Fires—Lab Tests and Reality

Fire is a very complicated process because it spreads. This means that materials in different places in the fire scene are usually in different stages of the fire. While materials near the point of ignition are in full combustion, those on the other side of the room may have reached decomposition, and those just outside the door are merely heating. As the fire spreads, the geography of the stages also spreads and the timing of the stages changes.

## THE LAG PHASE

After ignition, a fire may remain small for several minutes and grow slowly. This is called the **lag phase** of growth. During the lag phase, fuels are heated so that combustible vapors are formed. Most flaming occurs once these vapors are formed. After this initial lag time, the fire will grow exponentially. If the ignition occurs in an enclosed space, **flashover** may result after a few minutes. Flashover is the sudden bursting into flame of all the combustible surfaces in the room because the open flame has radiated enough heat to bring the vapors from all the materials to the crucial ignitable concentration in the air.

## FLAME SPREAD

The rate and direction of flame spread depend upon the placement and quantity/quality of fuel, air circulation, ambient (surrounding) temperature, moisture, and building design.

Fires spread rapidly and usually in an upward direction, because the heated gases expand and rise. However, if the fire has no fuel in the upper area of a room, it will slowly spread along fuel available in the downward direction. Fire beginning on the top floor of a multistory building will eventually spread to the floor below unless extinguished.

## THE SPREAD OF SMOKE AND GASES

Smoke and hot gases also spread upward and radially. If the fire becomes large enough to set up its own airflows (see Figure 2.1), the smoke and hot gases will travel on the convection currents, which are flows caused by a heat source—hot air rises, and cold descends. The smoke and hot gases travel much faster than the fire itself, and this is one of the reasons why victims generally die of smoke and fume poisoning before flames reach them.

A given volume of combustible solid material will produce a surprisingly large amount of smoke. Plastics are denser than natural polymers and are often fire-retarded, and they will produce denser and more voluminous smoke, unless the plastic is one of the few specially developed low-smoke varieties. The larger volume of dense smoke means that at a given distance from a fire, a given volume of plastic, especially fire-retarded plastic, will deliver a heavier smoke dose than the same volume of wood or cotton. In fires fueled by certain plastics, people situated at a great distance (300–400 feet) from the fire may die. The dense smoke causes

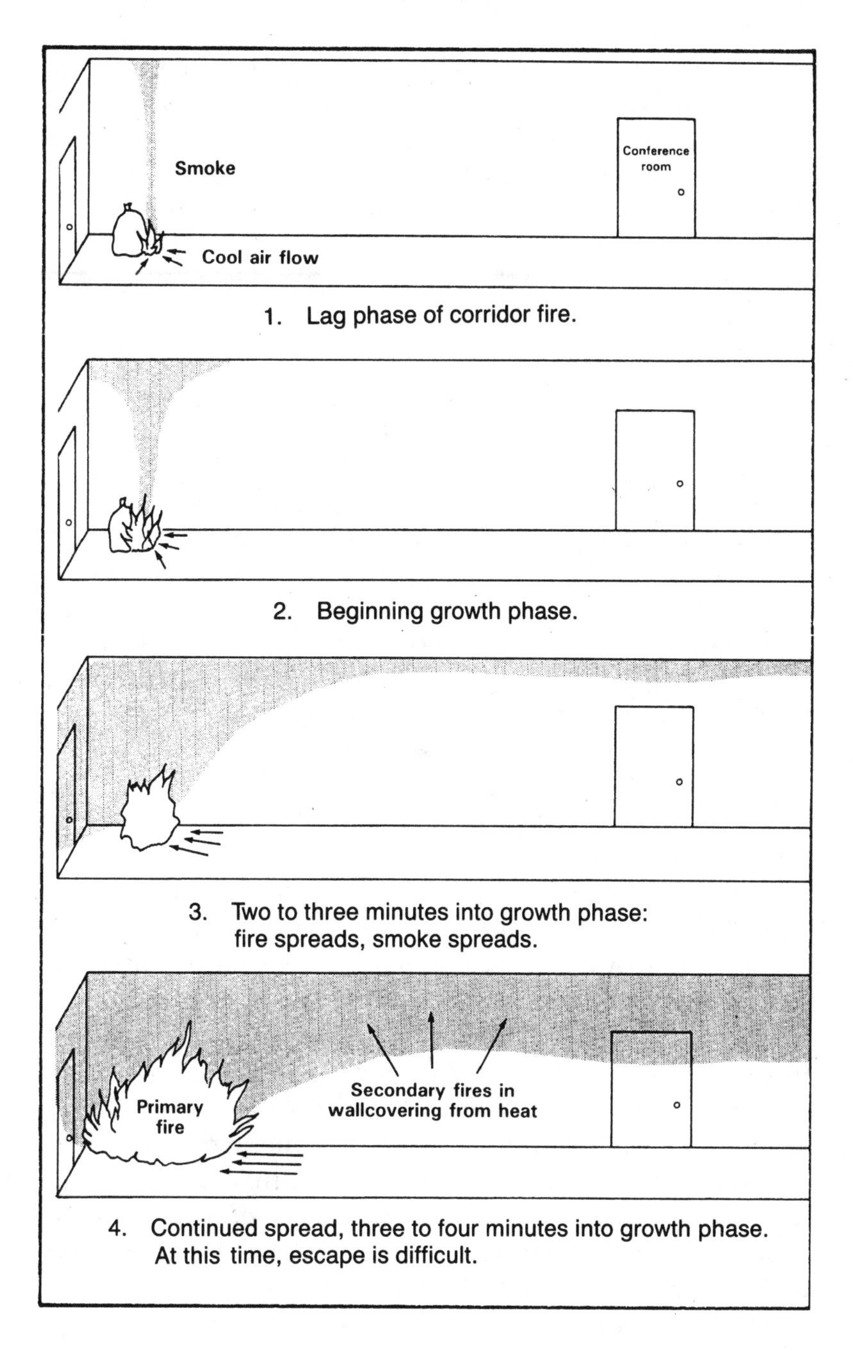


**Figure 2.1.** Growth of a Fire and Spread of Heat and Smoke.

visual obscuring farther from the fire, and causes people to become disoriented suddenly and without warning. The smoke and gas themselves can ignite.

Spread of the fire can be exacerbated by plastics in two other ways: the dripping down of burning plastic (the so-called flowing fire) and the sudden, large heat release that occurs when ignition takes place and the high energy and instability of the plastic's chemical bonds suddenly emerge.

Once they ignite, plastics tend to burn faster and hotter than natural polymers. Because plastics are petrochemicals (chemicals derived from petroleum or natural gas), they are generally high-energy materials.

Because smoke and fumes kill and injure people at a distance from the fire itself, the flow of air and gas assumes special importance in fire safety. The fire establishes its own circulation of atmosphere. Hot smoke and gases rise and travel along the ceiling, and cooler air rushes along the floor to the fire to fill the partial vacuum created by the rising, expanding, heated gases, as shown in Figure 2.1. The smoke and hot gases will rise from the fire floor to upper floors by stairwells, shafts, breaches in walls or between floors for building systems, and through the ventilation system. Prevailing outdoor winds can even blow vented smoke back into the open windows of a fire building's upper floors.

If the smoke and hot gases have no other exit, they will even travel to lower floors. This happens when pressure from the heated expanding cloud builds up, and the upward openings can't relieve that pressure sufficiently. The smoke will diffuse downward. Also, if the smoke and heated gases impinge upon a cool surface and are suddenly and dramatically cooled, they will drop down. This plummeting is especially characteristic of decomposition products that, at equal temperature, are heavier than air, such as hydrogen chloride (which PVC emits). Heat is the only reason for the rising hydrogen chloride plume from decomposing PVC. A low-lying "steam" or fog of acid gas is also a common emission of this plastic.

Many scientists want to predict detailed patterns of injury and death from a fire scenario of materials and building design. But transient environmental conditions that determine reactions such as diving smoke plumes make these prediction attempts a big waste of money. The knowledge of the quantity and types of combustibles that are present in a room or building is enough information to conclude whether or not smoke and fume deaths or injuries are probable in a fire.

In large buildings with air handling systems, smoke and fumes often spread far from the fire itself. The return air ducts will suck in the cloud of smoke and pass it to the cooler/heater, and from there it goes to the air supply ducts. The air supply ducts then vent it to the various habitable spaces in the building. Although much dilution of the smoke plume occurs during this trip around the air handling system, the dilution may not lower the airborne concentrations of chemicals to below injurious levels, especially if large quantities of plastics are decomposing.

In "tight" buildings, spread of high concentrations of decomposition/combustion products will be most hazardous. In these buildings the air recirculates and rarely if ever is diluted by outside air. As discussed in the previous chapter, "tight" buildings, even without a fire, often cause "office-worker illness" among people living or working in them. This condition is caused by the accumulation of irritants and organics in the recirculating air. In a fire, this everyday problem assumes grave proportions.

Although, theoretically, much dilution can occur as the smoke travels through the air handling system, don't bet on it. The cloud may move as a surprisingly cohesive unit and diffuse and dilute very slowly. It may be delivered in concentrated form to only a few of the possible recipient spaces. The design of the system and its mixing devices partially determine how much dilution takes place. These factors also determine the concentrations of toxic smoke to the habitable spaces and which habitable spaces in the building will receive the decomposition/combustion product.

A building can have an air duct smoke detector that can shut down the air handling system so that the spread of smoke via the system is prevented. But not many buildings have these devices, and some that have them don't maintain them in working order.

## FLAME SPREAD TESTS

Several independent organizations, such as the National Fire Protection Association (NFPA), Underwriters Laboratories (UL), National Bureau of Standards (NBS), and American Society for Testing and Materials (ASTM), have devised a variety of tests for flammability and flame spread. These ignitability tests range from simple attempts to evoke combustion with a standard bunsen burner held to the material for thirty seconds to complex delicate protocols. These delicate protocols include subjecting the material to controlled increases in temperature under strictly controlled conditions of airflow, humidity, and energy source. In some tests, the actual heat flux across the material is measured.

Flame spread tests generally measure how far a flame will travel along the sample within a standard duration under standard conditions. Again, a number of organizations have devised several different protocols. Some involve vertical orientation of material, because flames tend to move upward. Others place the material horizontally, especially floors and floor finishings, because that is how they are placed in normal use.

Although the standard-setting organizations acknowledge that these flammability and flame spread tests can't mimic conditions in a real fire, they also defend these tests because they yield valuable information about the relative behavior of the various materials in a set of controlled conditions. The tests yield information that is important to know before a material is marketed—the tendency of that material to ignite or support flame spread.

We must note, however, that these tests explore only the combustion stage. The generation of combustible gases in the decomposition stage is not usually measured in these tests, but it should be. Chapter One explains the dangers of the decomposition stage, and these dangers should not be ignored. The tendency of a material to spread secondary fires via clouds of combustible gases is usually not tested, either. The omission of these two areas is important. The way materials react with accelerants such as gasoline and kerosene needs exploration also, especially in view of the marketing of kerosene heaters and the storage of kerosene in polyethylene containers. Because of the popularity of kerosene heaters, more people are now storing kerosene in their homes.

## SMOKE GENERATION TESTS

Besides being tested for flammability and flame spread, materials are sometimes tested for smoke generation and obscuring potential. The two major ways of measuring smoke generation are (1) concentration of smoke particles per unit volume of air, after heating a given weight of material under standard conditions and (2) the degree to which smoke generated under standard conditions attenuates (obscures) a standard light source.

Some tests may actually track the weight lost by the sample during the test so that correlation with soot concentration or light attenuation can be made.

The degree of light attenuation by soot depends on both the number of soot particles in the air and the size distribution of these particles. If many small particles are present, this will cause more attenuation and obscurity than a few large particles, although the total weight of airborne soot may be equal for the two conditions.

Soot particle size and concentration determine several important safety features in a fire:

- Whether people can see to escape without becoming confused.

- Whether firefighters must take extra time to find the fire in the midst of thick smoke.
- The dosage of soot-bound products of decomposition and combustion received by people confronted with the smoke.

Materials show great differences in their smoke generation patterns. Materials that are fire-retarded often emit extremely dense, obscuring smoke because they don't simply burn to carbon dioxide. Fossil fuel-based materials often give off denser smoke than natural recent polymers (wood, cotton, linen, wool, hemp, leather) because the carbon in fossil fuels and their "children" is highly packed and the chemical bonds are high-energy compared with recently grown natural polymers.

## TESTING FOR CHEMICAL COMPONENTS OF SMOKE

The various organizations that design and approve standard tests for flammability, flame spread, and smoke generation never endorsed a standard test for exploration of chemical components of smoke and fumes. A large number of papers published in scientific journals report on the major and minor chemical constituents of fire gas and soot from various materials heated under a wide spectrum of conditions. Generally, these papers agree with each other as to the major chemicals. Most authors note the large quantities of hydrogen chloride emitted from PVC and the pattern of hydrogen cyanide emission from urethane foam.

It is when the authors describe the organic chemicals found in lesser concentrations but of great potential toxicological importance that differences become apparent. Hundreds of different types of organic chemicals have been listed as emitted by a given plastic. One of the trace constituents rarely described is PVC pyrolysis/combustion products in the dioxin/dibenzofuran family. The soot most commonly analyzed in these experiments is generated at high temperature in a *helium* atmosphere. Dioxin and dibenzofurans require presence of *oxygen* for formation. They are formed during cooling of the gases and soot. Thus, sampling of the hot soot right off the material, especially material burned in a helium atmosphere, precludes finding them.

There are influences other than airflow on the organic chemistry of the soot and gases. These influences include rate of temperature rise, additives in the material, presence of other materials decomposing or burning, humidity, and the geometry of the material (whether it is a thick solid, thin film, rolled up, stacked, etc.). The massively emitted products—those products that account for most of the weight lost by a plastic—pose the most immediate threat to life in a fire. However, the organics (carbon compounds) can account for 10–50 percent of the weight lost by the plastic and can influence the effects of the primary smoke chemicals. These organics may also cause long-lasting health problems to injured survivors.

In the last ten to twelve years, procedures for toxicity testing of fumes and smoke have greatly developed. In 1983, New York State's Department of State released a report done by Arthur D. Little Corporation. This report reviewed the fourteen major test procedures before focusing on the two widely accepted protocols, one developed by the National Bureau of Standards of the U. S. Department of Commerce, and the other by the University of Pittsburgh.

All the procedures reviewed in the report shared a number of features. The sample material was subjected to defined temperatures for a defined length of time; the fumes and smoke traveled through a tube to the animal exposure chamber; and the animals were never directly in the same chamber as the sample, nor were they subjected to heating.

The tests differed in many ways. The differences are listed below.

- In some, the materials were subjected to a fixed temperature or to a controlled rise in temperature.
- Fumes and smoke were moved by diffusion to the exposure chamber, or were carried by a recirculating airflow,

or were diluted and carried by a once-through airflow that was vented from the exposure chamber.

- Animals were sometimes exposed individually, at other times in a group.
- In some experiments, the animal's entire body was exposed to the fumes and smoke; in others, only the head was exposed.
- Sometimes the animals were restrained, and at other times they were on exercise devices.
- Different animal species were used.
- Different exposure times (fifteen minutes to four hours) were used.
- Post-exposure observation times ranged from minutes to three weeks.
- Different endpoints were recorded (time to stagger, time to collapse, time to death, number of deaths, weight of material causing 50 percent of the animals to die). An endpoint is the experimental result for which the test is performed.

Because of these different conditions and recorded endpoints, the relative toxicities of common materials varied from test to test. Years ago, when this testing first began, several astute researchers compared two methods for testing materials and pointed out to their colleagues how the different conditions used in the two methods led to different ranking of materials for relative toxicity. In general, plastics that emit acidic gases show greater toxicity in tests featuring rising temperatures and airflows.

## TWO MAJOR PROTOCOLS

There are two tests that are used most often and that are considered to be standard protocols. One of these tests is the **National Bureau of Standards test.** This test features a thermally static treatment of samples. A duplicate series of weights of a specific material is subjected to a temperature 25°C below ignition and to a temperature 25°C above ignition temperature. The cup furnace is preheated. (A cup furnace is simply a cup-shaped sample holder.) The test is static with respect to airflow also, and the smoke drifts passively into the animal exposure chamber.

The exposure chamber has a volume of 200 liters and renders a head-only exposure to as many as six male rats. Exposure for thirty minutes is followed by a fourteen-day observation period for surviving rats. Data recorded include gas analysis (carbon monoxide, carbon dioxide, and oxygen), chamber and furnace temperature, and thirty-minute and fourteen-day deaths. The major endpoint is the concentration of material causing 50 percent of the rats to die, known as the $LC_{50}$.

The other major protocol for combustion toxicity testing is the **University of Pittsburgh procedure.** In this procedure, the sample is exposed to heat in a furnace. The heat is increased at the rate of 20°C per minute. A total of twenty liters of air per minute dilutes the smoke and fumes and pushes them into the exposure chamber. The sample holder sits on a scale and weight losses of the sample through time are recorded.

Four male mice are given head-only exposures in the 2.2 liter chamber. Exposure time of thirty minutes is followed by twenty-four-hour observation. Data recorded include gas analysis, weight loss of the sample over time, temperatures of furnace and chamber, time to death, and $LC_{50}$. The university also does studies of the tissues of both dead and surviving animals to understand the nature of the injuries. Although the major endpoint is $LC_{50}$, time to death has also been emphasized as important.

## THE NEW YORK STATE RECOMMENDATION

The New York Department of State recommended that the Building and Fire Prevention Code Council adopt the University of Pittsburgh protocol and use it as a basis for generating a combustion toxicity data bank. This data bank would contain information on the toxicity of all construction and interior finishing materials. Product manufacturers and the real estate industry successfully blocked the implementation of this recommendation for several years. A series of objections created hurdles of delay and included far-fetched predictions of economic impact. The tests cost only $2,500 per product! Some industries spent many times that amount in entertainment alone to market a single product. The industry not only accused the state of heavy-handed regulation, but also had dire visions of being unable to market new products.

Some of the more plausible "reasonable" criticisms of this recommendation came from the mouths of industry-paid technical personnel. They said that:

- A laboratory test cannot recreate real fire conditions.
- The relative toxicity rating from a laboratory protocol cannot predict the toxic hazard of a material in a given fire.
- Carbon monoxide has always been the major fire toxicant, and everything organic gives off carbon monoxide and is therefore roughly similar in its fire toxicity.

Let us look at these plausible objections one by one, beginning with the last. Although nearly all organic materials, both natural and synthetic, emit carbon monoxide in a fire, many emit other products as well that heighten the toxicity of the smoke. In its campaign to confuse citizens, the Society of Plastics Industries (SPI) is fond of pointing to silk and wool as natural products that emit hydrogen cyanide, a statement that is inconsistent with their stand that the only important toxicant of everything organic is carbon monoxide, and that everything is similar in fire toxicity. As we have discussed before, many plastics begin decomposing before any carbon monoxide is released. These decomposition products tend to consist of acid gases, aldehydes, and other irritants. Although most fire fatality victims have high levels of blood carbon monoxide, evidence shows that carbon monoxide is not the only factor in many fire deaths and that the victims had been incapacitated by the irritating products of decomposition long before their blood carbon monoxide levels became fatal. Their tissues also showed the damage typical of irritant action. The evidence for rapid incapacitation is presented in the following chapters, which describe events in real fires.

The other two criticisms essentially accuse the lab tests of being lab tests. Most responsible manufacturers have no problem with flammability and flame spread tests, and this criticism would also apply to those if the wording were changed to fit the issues. Most manufacturers and regulators *want* information on the general flammability behavior of the products under standard test conditions. But the manufacturers, and some regulators, resist the acquisition of information presented to them on the products' general toxic behavior under standard test conditions. This refusal is inexcusable, since it is the toxicity of the smoke and fumes that kills the great majority of fire fatality victims, not the flames. Because of a double standard on testing for flammability and for toxicity, the establishment of a data base in New York State was delayed. See Chapter 10 for more about this.

Behind the plausible stories about economic impact and technical flaws looms the real reason for industrial "allergy" to the University of Pittsburgh protocol. The real reason is that many common plastics yield striking results when subjected to this protocol. They are much more toxic than rival materials on the market for the same uses. The square-inch area of wall covered by an $LC_{50}$ quantity of vinyl wallcovering is much less than that covered by an $LC_{50}$ quantity of wallpaper and much, much less than that covered by an $LC_{50}$ quantity of ordinary latex paint. Obviously, plastic plumbing pipe (PVC, ABS, or CPVC) shows much greater toxicity than steel, iron, copper, or glass plumbing pipe.

Food and drugs cannot be legally marketed without extensive, standardized toxicity and microbiological tests. Everyone acknowledges that even this testing cannot predict events that may occur in the real world, such as interaction between a drug and another source of chemicals such as another drug, pollutants, or a nontraditional diet. This gap is a signal that special testing is necessary when indicated, not that standardized screening tests should be done away with. The same philosophy should apply to the combustion toxicity situation.

The University of Pittsburgh protocol appeals to many fire scientists. It appeals because of its thermal and air dynamics, its individual exposure, and its method of tracking material weight loss. In real fires, products generally experience rising, rather than static, temperatures. Fires, by their heating nature, establish airflows and occur in buildings with either passive (relying on the building's structure), or active (fans) ventilation systems. The animals in the lab tests cannot huddle together and filter the smoke through each other's fur. In other tests where whole groups of animals are exposed without restraint, they put their noses in each other's fur and breathe "fur-filtered" smoke. Because they are restrained during the University of Pittsburgh test, with only their heads exposed, they cannot do this.

The degree of decomposition during the test can be correlated in time with such events as temperature and animal responses to the exposure. The records of the weight loss of the sample shows the course of decomposition, which can be followed along with the recorded furnace temperature and the recorded observations of the animals. Thus, at any point of the test, the furnace temperature, the weight loss, and the status of the animals are known. The progress of these three data sets can be followed and correlated.

The University of Pittsburgh protocol now generates combustion toxicity data on building materials for the New York State Building and Fire Prevention Code. This data bank should be part of our national fire prevention and control program.

*CHAPTER 3*

# The 1975 New York Telephone Exchange Fire

Around midnight on February 27, 1975, a fire broke out in the cables leading to a major New York Telephone switching station in lower Manhattan. By the time it finally died, this blaze consumed miles of cable; bent structural steel; spoiled concrete; put 700 firefighters to work, who used over 1,000 air canisters; injured 239 firefighters, officers, and support staff; and enveloped the neighborhood in a thick, acrid smoke plume, which sent hundreds of citizens to hospital emergency rooms. It was the largest fire in the Fire Department's history since the Depression Era's conflagrations of several blocks.

## THE BUILDING'S DESIGN

The switching center was (and still is) a drab looking gray concrete building, one block long, one-half block wide, and eleven stories high. Squatting at Second Avenue and Thirteenth Street, and overshadowed by Con Edison's "Tower of

Light" (Consolidated Edison supplies power for New York City), it had, until the fire, been ignored by the residents and workers of one of the most densely populated neighborhoods in America—the Lower East Side/East Village of Manhattan.

The firefighters didn't think much about it, either; they were being run ragged by the epidemic of fires in the Lower East Side and by the closing of fire companies. Rising workloads and falling fire-control resources were injuring and exhausting these men even before the fire. Between 1972 and the end of 1974, five of twenty-seven total companies in lower Manhattan had been closed. After the fire, the area lost two more companies and one man was taken from each remaining unit shift staffing level.

The cables leading to the switching center consisted of a mixture of old lead-covered dinosaurs and newer polyethylene-sheathed bundles of lines. The building and fire code allowed the use of flammable polyethylene sheathing outside the building. In the basement cable vault, the polyethylene and lead were joined to the required fire-retarded sheathed cables. The sheathing and insulation of individual wires inside the building consisted entirely of plasticized PVC: 40 percent phthalate plasticizer and 60 percent PVC and a small amount of additives (stabilizers, colorant, lubricant).

The basement vault resembled a sunken maze filled with rows and rows of cables. Although several doors to the vault opened on various sides, only one led to the ladder-way into the vault. Lighting in the vault was limited; the aisles between the cable banks were narrow; and, in general, conditions for mobility were very poor.

Risers (step-like parts of the cable support frame) led from the cable vault to the upper floors. The risers were in two raceways on opposite ends of the building, which served the first through fifth floors. The cables led up to floor five where they entered the distribution frames, and the structure of that floor was also mazelike with distribution frames. Between 200,000 and 250,000 pounds of plasticized PVC stretched from the vault and onto the distribution frames. The raceways had no firestops, no fire suppression system, and no state-of-the-art alarm system.

The windows of the building consisted of wire-glass, and on the first floor an additional layer of lexan (unbreakable transparent plastic) protected the wire-glass. Those are windows that will not break!

At this time, many AT&T switching centers had similar design and construction. Although some changes have been made, there are still fire problems in these switching centers around the country.

Bell Labs, the research institute of AT&T at the time of the fire, employed a combustion chemist/engineer by the name of Charles Bendersky. Months before the fire, Bendersky had circulated a report around the company about potential fire problems at the switching exchanges. He had evaluated the materials, especially the plasticized PVC cable sheathing and wire insulation, and the design of the installations. Based on the materials and the design, he developed a mathematical model that predicted events in a cable fire. The corporate executives at Bell and AT&T refused to act on the report and treated it as if it were one of the products of an academic pure research euphoria—a "high" scientists get from basic and theoretic research. But Bendersky knew better and tried several times for action on his findings. Unfortunately, no action was taken.

Why would Bendersky have known with certainty that disaster dwelt in those drab gray buildings? The reasons are very clear. They are listed below.

- The quantity and quality of the cable sheathing and wire insulation.

  PVC decomposes at the drop of a hat, and plasticized PVC, during decomposition, emits both HCl (hydrogen chloride, an acid) and plasticizer (a combustible gas).

- The PVC was installed vertically and without interruption in two raceways.

- There were no real fire control devices along the cabling.
- The building design was chimney-like.

The cables leading to the building, some of which were polyethylene, were quite flammable. Polyethylene is simply a long carbon chain with hydrogens on it. Classed as a polyolefin, it is closely related in chemical composition to candle wax (paraffin). We all know how candle wax burns, and polyethylene burns nicely also.

The complete materials-and-design picture of the Exchange can be summed up briefly. Highly flammable cables led to a subbasement vault that was full of cables both sheathed with and containing wires insulated with plastic that decomposed easily in heat. The decomposition products of this plastic were toxic and combustible. The indoor cables ran up two sides of the building to the fifth floor, and there was no firestop between floors and no extinguishment system in the vault or raceways.

Numerous other materials were present in the building including lead outdoor cable at the basement vault, PCBs in the transformer, some preserved wood, and the copper of the wire. Compared with the amount and toxic burden of the PVC present, however, these other factors are miniscule and of little importance in the course of the fire. Both Bendersky's model and the actual "experiment" showed that the placement of so much plasticized PVC into a structure of that chimney-like design overrode all other factors.

**THE FIRE**

On the night of the fire, the first alarm was turned in at eighteen minutes after midnight. The first arriving units saw no smoke outside, but were directed to the basement by two telephone company employees, who described heavy smoke. The smoke, indeed, was so thick that they couldn't locate the fire, and some smoke had spread throughout the building. Two telephone workers were also reported missing in the building. The firefighters began to search for the fire and for the missing people.

At 12:50 am, a second alarm was turned in. Several hose lines were stretched, and Rescue Company 1 made a forcible entry into the basement. Finally, they found the back door of the cable vault. On opening the door, they found very heavy smoke and heat. The firefighters from Engine 14 and Rescue 1 backed out and tackled the center door to the vault and found a large fire beginning to creep out of the vault through a hole in the ceiling.

Because of the density of the smoke, everyone wore a lifeline. The rope was not misnamed—anyone going into that smoke would have been lost. The vault was built like a maze, and visibility was zero. At this point, the problem of airpack depletion became evident. Engine 14 backed out because of depletion and were relieved by Rescue 1. Eventually, Rescue 1 also had to be relieved because of airpack depletion.

The firefighters tried to extinguish the fire with foam (it couldn't reach the fire because of the mazelike structure) and hoses. They found that fire could be put out only in one "lane" of the vault at a time. This is how control of the fire was gained—one section at a time, with the men crawling through the cabling.

When all the fire in the vault died, the heat remained and continued to cook the cable. The cooking cable gave off gases that accumulated under the vault ceiling. The flames in the vault were extinguished at about 3:30 a.m., but the gases touched a hot surface at about 4:00 a.m. and exploded into a conflagration that raced up the cabling to the floors above.

During the first extinguishment of the vault fire, the building resembled a seething anthill. Ladder companies struggled with the wire-glass windows and the lexan on the lower floors to vent the smoke and gases. Other companies searched for the two "missing" telephone employees—as it turned out, they weren't in the building at all. When the fire moved up to the first and second floors, other companies were called.

Firefighters on the first floor also found themselves in a maze of metal and glass partitions, distribution frames, and high ceilings. These combined with the dense smoke to initially hide the extension of the fire from the basement. Even during the first fire, the officers knew that the interacting factors of materials and building design were extraordinary and potentially overwhelming.

The fears of the firefighters began to express themselves in various ways. Some were afraid that the fire would spread to other buildings. Small problems were inflated to large ones by the firefighters' emotions. Visions of civilians possibly being trapped in the building revitalized the search activities. One small event that was blown out of proportion was when a firefighter lost his helmet in the midst of the darkness and acid smoke. It was duly noted that the proper papers would be filed to report this incident.

All of the above are standard operating procedures for large-building fires; officers *do* think about spread to other buildings, they *are* charged with keeping track of the Department's equipment, such as helmets, and they *must* order searches for people who may be trapped. However, the fire report gently hints that a panic was growing, and that it resulted in a clinging to routine and to familiar decisions. The bureaucracy of the firefighter's lost helmet leaps to the eye in the midst of chaos: "A helmet was reported lost and not recovered by Fireman 1st Grade———, Rescue 3. Proper reports are being processed."

In the half hour following the extinguishment of the vault fire, everyone thought that the fire was under control and that only a little mop up of the upper floors held them in the building. Then an explosion caught them by surprise—as did the fire from it.

The explosion was exactly that—BOOM! One firefighter wrote this description of it: "I was on the sidewalk just outside the vault, tending hoses. We were going to start packing up. All of a sudden, there was this big bang, and I'm knocked down to the sidewalk. Suddenly, I'm sitting down. That hurt."

The building conditions worsened considerably with the second fire ignition. There was much more dense black smoke, even in the upper floors, and this smoke was more acidic and allowed no visibility. The firefighters withdrew from the basement and lower floors for a short time to reassess the situation and to regroup. Airpacks needed changing as well.

The first firefight was directed by Division 1. Even during the first fire, the smoke issuing from the building led the Division Chief to order evacuation of the area around the building, including the Manhattan Eye and Ear Hospital. With the listing of this necessity, the fire report indicates the extraordinary conditions prevailing even during the first, smaller fire.

The second fire required thirteen additional hours to control. Three hundred firefighters worked on-site during the most intense control effort, and a total of 700 firefighters worked the fire from start to finish. The average fire in New York requires twenty minutes work from four units. Firefighters reported presence at the fire scene for durations ranging from one to twelve hours, with most serving four to eight hours. Over 1,000 air canisters and miles of hose were used.

Ambulances and vans shuttled between the fire scene and hospitals. Fire Department ambulances stayed at the fire scene and were turned into examination rooms staffed by Medical Division doctors and paramedics. Nearly all support units were activated: satellites, mask service, foam, communications, the superpumper, and designated command post.

The Chief of Department/Commissioner arrived on the scene with a bunch of headquarters officers including the chief of training, chief of communications, Manhattan Borough Command, and the chief of planning and operations research. Even Mayor Beame put in an appearance.

Although the heat from the burning PVC was horrendous (steel beams buckled on the third floor and rendered that floor structurally unsound), the worst factor in this fire was the smoke. Fire Department ambulance drivers and doctors stationed two blocks away from the building became sick from smoke inhalation. People who lived and worked for blocks around flooded the local emergency rooms. Throughout the fire, the firefighters engaged in disciplined firefighting teamwork.

In all, the units made three major attempts to gain entrance to the fire floors. The first was between 5:00 and 7:00 a.m., the second at about 9:00 a.m., and the third at 11:30 a.m. The third succeeded. The heat, the hostility of the fire, and the density and acidity of the smoke repulsed the first two attempts. Tower ladders directed streams of water into the firefloors from the outside, but really could not reach the core of the various fires. The fire report states: "Stream penetration was impeded by rows of cable frames, catwalks, columns, partitions, delayed window ventilation, low windows in relation to ceiling height. (This factor also impeded ventilation at ceiling level.)" Diminished fuel availability probably accounted for the eventual die-back of the flames. The fire had been burning for eleven hours when the third entry attempt succeeded.

Firefighters preparing for a fire set out to defend themselves against an invading force of known size and nature. But at this fire, they confronted the unknown, the unexpected, a phenomenon for which they were unprepared, an enemy whose dimensions and weapons mocked their technical art and strategy. They could not even find the enemy's strongholds, not for a long time! In the midst of the blackness, the flames would suddenly flash. Multiple lines were manned by two engine companies operating alternately. Companies were stationed on the fourth and fifth floors to contain the horizontal spread of the fire from the floors below. During the last third of the sixteen-hour ordeal, the men who were stationed on the upper floors faced heavy smoke that had accumulated and risen from the floors below.

The PVC in the cables was being "cooked," and as a result other minor flashovers occurred besides the major one that reignited the primary fire. On all the lower floors (1–5), the firefighters encountered scattered fires in the cables. On other floors such as the ground floor and fifth floor, where the cables left the risers and joined the distribution frames, the fires could have been anywhere, on the risers or on the distribution frames.

Engrossed in venting attempts or in fighting the fires, the men often failed to realize that their air canisters were depleted. Many of them had to breathe the thick acidic smoke and fumes when the airpacks emptied. The fire report understated the problem.

> Scott cylinders were occasionally overextended by members desiring to complete search for victims due to size of building, limited time on cylinder, time required for personal withdrawal, and also by members attempting to hold ground gained while awaiting relief in early stages of operations in cable vault.

The later survey revealed that whole teams routinely found themselves with depleted airpacks on the upper floors as well as in the cable vault.

A few men panicked, became disoriented, and had to be controlled by their comrades. By far, the great majority kept self- and company-discipline, even though they faced both sensory assault and respiratory punishment of an unprecedented nature and degree. The fire fighters and officers in the building actually subjected to the undiluted poisonous dragon's breath showed greater calm, judgment, and dignity than the desk officers and doctors outside.

## THE RESULTS

Table 3.1 lists the companies that worked the fire, a total of 75—the equivalent of twelve to fifteen alarm assignments. The list is slightly misleading in that several of these companies worked two or three shifts in succession. A total of 700 firefighters served at this fire. Three hundred of them worked simultaneously at the height of the fire after the reignition. Again, the numbers are somewhat misleading be-

**Table 3.1.** Firefighting Resources Dispatched to the Telephone Fire.

| Borough | Engine Companies Responding | Ladders Responding | Rescue Squads Responding |
|---|---|---|---|
| Manhattan | 25 | 14 | 2 |
| Brooklyn | 4 | 4 | 2 |
| Queens | 1 | 1 | |
| Bronx | 1 | | |
| TOTAL | 31 | 19 | 4 |

| Others (From all boroughs) | | |
|---|---|---|
| | 13 | battalion chiefs and their aides |
| | 2 | satellite units (spotlights) |
| | 2 | ambulances with paramedics |
| | 2 | medical cars with physicians |
| | 1 | foam unit |
| | 1 | superpumper |

Many of these units stayed at the fire for more than an entire shift. Manhattan had 44 engines and 33 ladders total and thus had only 43 percent of its engines and 58 percent of its ladders available to respond to all the other alarms of the borough. Certain areas of Manhattan such as Lower Manhattan were nearly stripped of available companies for hours. Under these circumstances, other fires can become larger and do more damage than necessary because of the late and inadequate response by the remaining available firefighting units.

cause many of the men worked more than eight hours on the scene. If we look at the question of resources and time, we must imagine a normal New York fire. In 1975, the normal New York fire received a response of four companies and a battalion chief (a total of twenty-two men) and was worked for twenty minutes. The New York Telephone fire, in terms of companies, was the equivalent of about 20 normal fires, and when the shifts and time are considered, forty-eight normal fires. The discrepancy between the time and the number of company shifts shows that even with the huge number of men and companies, this was a severely underserviced fire. The number of companies that could be sent without stripping the whole city of fire service had been greatly diminished by the company eliminations of 1972–1974.

## FIREFIGHTER INJURIES

Firefighter injuries are described below. Keep in mind that all the companies listed were being served by firefighters who were unable to work up to their peak performance because of these injuries. These, additionally, were the firehouses to which contaminated gear was returned and stored and where men lived with the stored contaminated gear. In 1975, most people weren't concerned with or aware of the dangers of dioxin or chlorinated hydrocarbons. Firefighters considered soot to be an annoyance, something that gave them the world's worst case of ring-around-the-collar. The awareness of how plastics act in fire changed all that.

The official Fire Department report lists 239 FDNY employees injured during the fire. This number can't possibly represent anything even close to the census of the injured men. Several men did not report sick—perhaps they wanted to perpetuate their gung-ho, macho image. One of these deniers, a lieutenant, died of a heart attack in early March, less than two weeks after the telephone exchange fire. His heart attack was suffered at a multiple alarm blaze. His autopsy revealed older, heavy deposits of greasy soot that had eaten its way completely through the lung to the pleural side of the lung. At the time of his death, he still had lung edema, and he had dead patches in the lung.

An undercount also probably occurred when men became ill 24–72 hours after the fire but never mentioned to their doctors that they were involved in fighting the fire. Delayed symptoms from inhalation of smoke from PVC or Teflon sometimes resemble flu, and firefighters may not have connected their "flu" with this fire.

In any case, we know that at least one-third of the FDNY staff active at the fire was injured. The great majority of these injuries involved smoke inhalation. Even medical personnel stationed two blocks from the fire were affected.

A later survey of the injured firefighters allowed a more detailed classification of their immediate symptoms than listed in the official fire report (see Table 3.2). The picture emerges of teams of men trying to control a many-headed chemical Hydra while they also controlled their own deteriorating physical and neuropsychological conditions. These conditions included acid-burned respiratory tracts, eyes, and skin, inability to get enough oxygen because of lung damage; loss of control over limbs; impairment of the whole perception process; nausea and feelings of weakness and exhaustion; and confusion and disorientation.

The symptoms arose in a typical sequence. First, the men were vaguely aware of irritation of the nose, throat, eyes, and

**Table 3.2** Immediate Symptoms of the Firefighters.

| Injury | Percent Affected |
|---|---|
| Sore throat, irritated eyes, dizziness, aching nostrils, confusion, weakness, and exhaustion | Over 50% |
| Chest pains, nausea, chest congestion, and headache | 35–50% |
| Irritated skin and faintness | 20–30% |
| Loss of control of arms and/or legs | 10–20% |

skin. Then the burning sensation extended down the throat to the larynx and became much worse. In time, the burning sensation extended to behind the breastbone, and coughing began. Frontal headache, different from the typical carbon monoxide headache, arose about the same time as the burning in the chest.

Later, the men felt tight of chest, as if they couldn't breathe deeply enough. Some lost a degree of control over their limbs and had "the shakes," especially in the legs. Their grip on equipment was weaker, also. Some experienced confusion in thinking; others experienced very slow thinking, with great effort. At this point many also suffered from dizziness and nausea, a feeling of drunkenness or motion sickness.

The actual timetable of symptoms cannot be found from the survey answers. It is known that for some, all these symptoms appeared within a half an hour after they began working at the fire. Others were so intent on their jobs that they became aware of their symptoms only after they were quite ill and into the chest-tightened stage.

The remarkable fact about all this illness is that most of the men continued to obey orders and work even if they had to take three times longer than usual to accomplish something. Whole teams were very ill, but everyone accepted the illness as simply part of the job. It was normal to be ill at this fire. When everyone has smallpox, the unmarked man is abnormal.

The survey that was taken five years after the fire revealed that recovery from the injuries took months. Some never recovered. Table 3.3 shows the injuries still present in the period two weeks to three months after the fire, and Table 3.4 shows those present six months or more after the fire. As you can see, although some of the complaints are the same, changes took place over time.

Some of the critics of the survey pointed out that not all the firefighters listed as injured on the fire report sent in answered survey forms. The response was over 60 percent. It is possible that firefighters who retained their injuries over a

**Table 3.3.** Intermediate Time Symptoms.

| Symptoms | Percent Affected |
|---|---|
| *Respiratory:* | |
| Chest congestion | 51.16% |
| Chronic cough | 22.09% |
| Sore throat | 18.50% |
| Sore chest | 9.30% |
| Hoarseness, wheezing, allergy to smoke, difficulty breathing, irritated nasal membranes, shortness of breath | Less than 5.00% |
| *Neurological:* | |
| Muscular weakness | 19.77% |
| Impaired smell/taste | 16.28% |
| Increased irritability | 10.47% |
| Headaches | 10.47% |
| Perception difficulty | 6.96% |
| Confusion, anxiety, numbness of extremities | Less than 5.00% |
| *Miscellaneous Symptoms:* | |
| Heart trouble, irritated eyes, irritated skin | 2 complaints each |
| Chills, sinus trouble, weight loss, bowel problem, nausea, head congestion | 1 complaint each |
| *General Well-Being* | |
| Fatigue | 5.81% |
| Impaired Endurance | 3.49% |

long time tended to send in their answered forms. However, several firefighters who were sick and came to our attention never sent in completed forms. Two of these men died of rare

**Table 3.4.** Prevalence of Long-Term Effects.

| Injury | Percent Affected |
|---|---|
| *Respiratory:* | |
| Impaired disease resistance | 37.5% |
| Coughing | 33.3% |
| Hoarseness | 23.61% |
| Shortness of breath | 9.72% |
| "Lung function" or pain | 15.8% |
| Chest congestion | 9.72% |
| Sensitivity to smoke | 11.11% |
| Sinus or nasal drip | 6.94% |
| Repetitive bronchitis | 8.33% |
| Sore throat | 8.33% |
| Asthma | 6.94% |
| Allergy, unspecified upper respiratory problem | Less than 5.00% |
| *Growths* (epidermal or membrane lining) | 13.89% |
| *Heart:* | |
| Palpitations, acute myocardial infarcation, prolapsed mitral valve, enlarged heart, unspecified damage | 8.33% |
| Headaches, perception difficulty | 4 individual complaints |
| Fatigue, kidney-urinary tract | 3 individual complaints |
| Weakness, pancreatitis/diabetes | 2 individual complaints |
| Elevated blood count, elevated bilirubin, high blood pressure, gall bladder deterioration, irritation of hemorrhoid, irritated eyes, convulsive seizures | 1 individual complaint |

cancers—one from brain tumor and one from liver cancer. A third was the only firefighter who sued, because his lungs were so damaged that at the time of his court date, they were functioning at only 50 percent of what is considered normal for his size and age. The survey forms, in all likelihood, are a good index, a good sample of the fate of the men injured at the fire, including those who didn't show up on the official list attached to the fire report.

Before we examine the injuries in detail, let's look at the characteristics of these men to see what they have in common. They were generally young men. The average age was about forty, and very few were over fifty-five. Because of the company closings and layoffs of men with less seniority, there were also few under twenty-eight. A prime group of firefighters were at this fire: healthy, young, but experienced and competent.

They showed a wide range in height (from five-foot-seven to six-foot-six) and weight (140 to 266 pounds). Only 23 percent of them were smokers. Some were joggers. As a group, they showed a higher level of health consciousness than the average group of workers.

Who tended to retain permanent injuries? Those over thirty-five years of age and those who had served more than ten years as firefighters. In fact, of those who had served more than fifteen years, 80 percent retained permanent injuries. We interpret this to mean that years of previous exposures to smoke had affected the ability of recovery in this group and had initially give them injuries which this exposure greatly exacerbated. The Harvard School of Public Health had studied Boston firefighters and found that lung function deterioration depended on both total number of fires fought in a career and the amount of exposures that were especially arduous.

Of the 113 respondents, 72 (63.7 percent) complained of persistent or permanent effects. This category is defined as injury present from six months after the fire through the time of responding to the survey, or recurring several times between the fire and the time of responding, or a present impairment arising out of an effect of the fire. For example, many men complained of being hoarse from the time of the fire through the time of responding to the survey. This condition became progressively worse in two men, who went to their doctors and were found to have vocal cord lesions. Impaired disease resistance, coughing, sensitivity to smoke, lung function problems/shortness of breath, and repetitive bronchitis were common complaints on the survey forms.

The two most serious complaints were heart trouble and the cases of vocal cord lesions. The high prevalence of hoarseness indicated potential trouble, and indeed, when the later cancer survey was conducted, the problem had grown.

Besides the injuries that were directly related to the tissue damage sustained during the fire, the cancer pattern in the cohort of 700 men differs from that of the fire department uniformed staff as a whole. Table 3.5 lists the six cancer cases found among the telephone fire cohort by late 1981. The following points must be kept in mind:

- Two-thirds of the cases occurred in men under forty-five years of age, whereas in the fire department as a whole, thirty-five out of the sixty-two cases occurred in men over forty-five.
- Three of the four cases of laryngeal/throat cancer in the Department came from the telephone cohort.

**Table 3.5.** Cancer Cases Among New York Telephone Fire Firefighters.

| Site | Age[1] | Years Service | Year Detected |
|---|---|---|---|
| Larynx | 38 | 17 | 1978 |
| Larynx | 37 | 12 | 1978 |
| Liver[2] | 37 | 14 | 1980 (fatal) |
| Skin | 41 | 14 | 1979 |
| Throat | 51 | 28 | 1979 (fatal 1980) |
| Brain | 53 | 24 | 1980 |

[1]Age at which cancer was detected.
[2]Same man had cancer of colon in 1976, was operated on, and returned to duty.

- Laryngeal/throat cancers constituted only 6.35 percent of the cases in the Department, but 50 percent of the cases of the telephone cohort.

- Although the cancer incidence of the cohort is only slightly higher than that of the department, the age-specific incidences are much higher.

After the cancer survey was conducted, one other firefighter who did not respond to the survey supplied information that he had been operated on for laryngeal malignancies during the study period. Also, a doctor who had served at the medical van contracted cancer of the esophagus. Clearly this cohort has excessive risk of laryngeal/throat cancer.

The injuries retained by these men and the apparent shift in cancer pattern should be no surprise when we remember the major chemicals, besides carbon monoxide, that exist in the fumes and smoke of PVC. Present are high concentrations of acid both as a gas and bound to the soot, benzene, chlorinated dioxins, chlorinated furans, and a mix of hydrocarbons. The acid damages tissue, and the organic chemicals initiate and/or promote cancer.

After the results of the survey were published in 1981 and 1982, the fire department itself began to monitor the group that had fought the fire. Without informing the men or the union, the department accumulated inactivity data. The data included the proportion of men who retired, what type of retirement (line of duty disability, ordinary disability, or simple retirement), and deaths. Rumor of the report circulated among the officers and finally reached Tom Gates, then Sergeant at Arms of the Uniformed Firefighters Association. When Gates obtained a copy of the report, he decided that the whole story had to be told publicly. *New York Daily News* reporters Don Singleton and Vincent Lee compiled the results of the injury survey, the cancer study, and the inactivity tally into one large feature story. When all the data appeared in one place, it was evident that an occupational health disaster had occurred. Because of the delayed and unusual symptoms, the firefighters had not connected their health problems with the fire. Some had even said that they thought they were simply getting old (and these were young men who were thirty-eight to forty-four years old). Others just thought that this happened to all firefighters.

The battle for awareness and justice for the injured men is far from over. The fire department still maintains that many of them cannot retire under line-of-duty disability because their injuries aren't the conventional firefighters' lung and heart problems, although some have been able to retire because of their lung and heart problems. The cancer cases, however, are being classified as non-line-of-duty. Those victims with upper respiratory tract damage, such as laryngeal cancer or scars, are likewise not permitted the proper classification.

Only recently has the New York State statute of limitations for victims of toxic exposures changed to allow the injured men to sue parties responsible for their exposure. The law had previously limited suits to those filed within three years of the exposure event. Now it allows suits filed within three years of discovery of the illness and its link to the exposure event. For five chemicals, it provides a time window of one year from the time the law became effective for currently time-barred victims; PVC is one of the five chemicals.

## EVALUATING MEDICAL SERVICE AT THE FIRE SCENE

The Medical Division of the fire department can be blamed, in part, for the failure of the firefighters to link their illnesses with their exposure to smoke at the Telephone Exchange fire. The Medical Division neglected to monitor the cohort systematically to reveal any long-term health effects. It appears that they also failed to document the immediate symptoms properly. Very early in fire operations, the extraordinary nature of the smoke and its effects on the firefighters must have been obvious. In such circumstances, ethical doctors keep minutely detailed records, because the future health of the patients may depend on the observations recorded during and immediately after the exposure.

Medical service at the fire scene was inconsistent and suffered from assembly line rush. Most of the respondents complained that they were given either no examination at all or a cursory visual one. Many were told to take the rest of their shift off and to report to their next shift, although the doctor could not possibly have had any idea of whether they should indeed report to work so soon. Some men were given one or two hours of rest and rehabilitation and sent back to work; others received their first medical attention in the hospital, but were given no time off; still others dug their heels in and were given time off.

The doctors tried desperately to cling to such routines as taking blood pressures or listening to lungs. Whether doctors were motivated by panic or by deliberate decision to keep the men working, whether they were physically able to or not, a combined serious lapse of technical competence and ethics pervaded and subverted the on-the-scene medical service.

A further gap in the medical service stemmed from lack of coordination between fire service and Health Department. No working medical system was set up. Communications, public information, and many other support systems were set up efficiently during this fire. But firefighter injury appears to have had low priority. The miserable doctor/patient ratio generated at this fire scene appeared again in several subsequent large plastics fires. This fire is the model for the strengths and weaknesses of not only the New York Fire Department in fighting these fires, but probably for most other large American cities.

Firefighters needed medical attention. Citizens also needed medical attention. The emergency rooms of the local hospitals were filled with neighborhood residents and workers complaining of choking and coughing, and irritated eyes, noses, throats, and chests. No study has ever been conducted on the aftermath of these injuries. We do know that a resident died of liver cancer seven years after the fire, and that several people called the union about contracting asthma right after the fire. The dimensions of this public health impact will probably never be defined.

## WHAT CAN BE LEARNED FROM THIS FIRE?

What are some of the facts we have learned from the Telephone Exchange fire of 1975? Let us simply list them.

- The outcome of a fire largely depends on the quality and quantity of the organic materials present and the structural design. Mixed synthetic fuel loads pose special problems.
- PVC in the stage of decomposition and combustion can deliver an acute dose of toxicants which results in permanent serious injury and even delayed fatalities.
- Although the concentrated cloud of acid poses the most immediate life and health threat, other chemicals, especially chlorinated hydrocarbons, can cause or contribute to serious chronic health problems.
- If the local building code allows large quantities of PVC in a building, the fire department and other city agencies must budget and plan for major disaster. These plans should include sophisticated short-term and long-term medical care for hundreds of people, both firefighters and civilians.
- All responsible parties failed to take the proper preventative measures, from the corporation that supplied the PVC, to Western Electric (the AT&T manufacturing affiliate), to the various city agencies, the National Bureau of Standards (NBS), the Consumer Product Safety Commission (CPSC), the NFPA, and the medical professional societies. All of these parties had prior knowledge of the problems with PVC in fires and failed to take action.

- Human reactions to the disaster were consistent with other types of major catastrophes. The reactions included denial and clinging to routine; anger, guilt and self-blame; and recurrent anxiety.

- Other neuropsychological reactions appeared related to the chemical exposure, especially a spectrum of emotional and motor affective disorders; weaknesses in the limbs, appetite changes, sleep pattern changes, depression and irritability, and hyperreactivity.

It is hoped that the consequences of this fire will lead to changes in the construction and layout of other buildings of this type. But even with improvements in construction and layout, the danger will not be completely eliminated. Synthetics are too pervasive in our world; we can never be too comfortable or confident about our safety. The Dragon lurks in the most unexpected places.

*CHAPTER 4*

# The Younkers Brothers Department Store Fire

The danger of the fires discussed in this book is their speed—not necessarily the speed of the flames, but the speed at which the smoke becomes incapacitating. On November 5, 1978, a fire in the Younkers Brothers Department Store produced a thick, black curtain of smoke that killed quickly. The toxic smoke was traced to the polyvinyl chloride (PVC) wire insulation in the electrical system of the building.

The store was located in a mall in West Des Moines, Iowa. Twenty-two people were present that Sunday morning; ten died, and at least four were injured. The first sign of fire was a low-energy explosion that occurred in the ceiling of the second floor, knocking down ceiling tiles. Immediately, a black curtain of smoke descended from the second floor ceiling in the southeast corner of the store, incapacitating all who came into contact with it almost instantaneously. The smoke rolled through the store, and those who could fled the dense blackness. Those who couldn't died. Survivors later said that the smoke was so black and thick that it looked like rolling death.

The heat from the fire traveled through the air ducts in the store and set secondary fires, but this was after the initial visitation. Outside, witnesses saw smoke pouring from the southeast corner. At first the smoke was white and hazy, then it became black and thick. The smoke was so dense that arriving firefighters had to turn on the headlights of their rigs while they were still approaching the parking lot from the highway. In trying to get into the store, their hand-held flashlamps proved useless against the pervasive darkness of the smoke.

## THE DESIGN OF THE BUILDING

The Younkers Brothers Department Store in Merle Hay Mall was originally built in 1958 as a two-story concrete building. Two additions were made, a one-story extension in 1963 and a two-story addition to the east side in 1967. The materials used were mainly steel, concrete, and other noncombustibles.

The store measured 345 by 200 feet and had exits on all sides. All of the people present in the store had worked there for at least several weeks, and most of them for over a year, so they knew the position of the exits. Although the NFPA attributes some loss of life to lack of sprinklers, all who survived had left the store before the habitable space was involved in flames. It is true that the store had no sprinklers; but sprinklers, as normally installed to protect only the below-ceiling habitable space, would not have saved even one life or prevented one injury in this fire.

The mall itself was a typical enclosed structure with a glass and metal roof. Hydrants had been placed in strategic spots around the mall, and water supply was not a problem, per se, in this fire.

Firefighters who were battling the fire noted that the roof contained smoke that was unusually dense and voluminous. Early in fire control operations, the skylight nearest the burning store was broken by firefighters for venting of the smoke.

## EYEWITNESS ACCOUNTS OF THE FIRE

After the fire, survivors were interviewed by fire department personnel, and later they gave depositions in the litigation proceedings. Many of the responding firefighters and officers wrote individual reports containing their observations. Store personnel described the extent and type of damage. A large number of residents and workers witnessed the event from outside the store, and one even took photos. The fire department staff also took a large number of photos of both the inside and outside of the store after the fire. In addition to all of the above, a small amount of physical evidence remained available for analysis. This evidence consisted of soot on the piece of clothing of a survivor who escaped very early in the fire, medical records of injured survivors, partial autopsies of all of the victims, and a second autopsy of one of the fatality victims whose family had requested exhumation and re-autopsy.

From all of these lines of evidence, a consistent narrative of this event emerges. The steps of the survivors could be retraced, and placement of their observations in time and space was possible. Through their eyes, the movements of the fatality victims could also be reconstructed. We know who was doing what and where they were at the time the ceiling tiles fell and the black curtain of smoke invaded the store.

The Des Moines Fire Department investigated the fire by analyzing physical evidence, interviewing survivors and other witnesses, and collecting the reports of many firefighters who worked the fire. The firefighters recorded a number of unusual observations, beginning while they were still approaching the mall. Several of them made remarks about the volume and density of the smoke and how the smoke caused traffic to slow down because of visibility. It is recorded that just "north of Urbandale the smoke was influencing traffic conditions." The fact that the smoke made visibility poor was a recurring one in the reports. At the parking lot, rig drivers had to turn on their headlights to see, and even then they had trouble avoiding thin objects like lightpoles. The firefighters could not see in the eastern half of the store because of the smoke density, and their powerful handlights did not help: "This atmosphere was very hot and the smoke so dense hand-lites [*sic*] were almost of no effect." Several firefighters noted the acidity of the smoke as well as its blackness and density:

> Lt.— ordered me to get a light plant, cord and large smoke ejector into mall corridor on the north side of the building. This corridor was filled with thick, acrid, black smoke and before putting the smoke ejector in position it was necessary to get air masks on.

The men who entered the building from the east doors described a green flame that they could not put out. They doused the flame with water several times, but it kept springing back to life. Because it was so persistent, they thought that it was a natural gas leak fire.

Although the firefighters said that there were small fires everywhere by the time they entered the building, they also concluded that these fires were secondary, and that the main body of the fire and smoke was on the second floor. All the raging heat and the origin of the dense black smoke was somewhere up the escalator on the second floor. But when they attempted to climb the escalator, they had to turn back time after time because the heat was so intense and the smoke so dense.

> We advanced inside maybe 80 feet or so and found an escalator going to the 2nd floor which we tried to go up and did get maybe halfway up, but then the heat got more intense and we just could not get all the way up there because of it. Then the bells on our masks started to ring and we had to retreat to come outside to change our air tanks. After changing tanks we went back inside again, but by this time the heat level had built up and we were not able to go back in as far as we were before.

Eight people who survived gave clear, consistent accounts of the event as they experienced it. Two other men who initially survived the fire will never be able to give their accounts. One died shortly after the fire of a mixture of cancers and lung disease, which may have influenced his ability to observe and to remember clearly, and the other man was mildly retarded and gave confused and inconsistent statements which showed that he was disoriented and could not clearly express or possibly remember what he did and what happened to him.

The following are quotes from fire department interviews of survivors.

> First we saw it coming from the east to us, so we tried to go north but then we saw the smoke from the south to us and by then we thought we were trapped. There isn't a door here. You can go out that way and this way. So we saw the smoke from both places, I opened the door here, we thought that was a way out. As soon as I opened it, I saw the room was completely filled with smoke too.

> I was facing south and talking to S—— and the wind came through that hallway. S—— turned around and I saw smoke drifting up toward the ceiling. Just thick smoke. We couldn't see anything, Just thick black smoke all over. It was like walking into a fog and losing everything. You couldn't see anything in front of you anywhere.

> Q. Was the smoke very thick at this time?
>
> A. It was fairly thick, 'cause I lost M—— after a couple of steps.
>
> Q. Was there any way you could tell where it was coming from? or going to? or could you tell?
>
> A. It was just there like a wall.

> I stood up at about 9:15–9:30 a.m. when I saw the doors give a little bit. These were doors to the west of me. I opened the doors and saw the hall filled with black smoke that looked like a cyclone.

Outside the building, people who lived and worked nearby

or who happened to pass the store at crucial moments also added their observations. These observations added to the understanding of what was happening in early stages of the fire, in the critical moments that determined who would survive and who would die.

People who were eating and working at the doughnut shop across the road from the store turned in the first alarm. They saw a stream of thin white, hazy smoke pluming out from the southeast corner of the store. Gradually, the smoke turned black.

A professional photographer who lived in the neighborhood ran to the parking lot with her camera when the first fire vehicles arrived. She took photos of the smoke plume described above coming from one corner of the building. She also took pictures that documented the damage that was done by the fire and explosion on the outside and the damage the firefighters had to inflict as they tried to vent the building.

## THE VICTIMS

A young man and his father happened to drive up to the store just after the explosion. They saw two men trapped between two sets of glass doors. The passers-by broke one of the glass doors with a tire iron and let them out. The firefighters had arrived by then and saw one man who had a reddened face and soot all over him, and who was coughing. This man was taken to the hospital. The two who used the tire iron said that the hospitalized man was covered with black soot, coughing, and gasping. The other man looked fairly clean, but was frightened.

Figure 4.1 shows where the fatality victims died. All were on the second floor at the time of death, and eight of the ten were in the eastern half of the second floor. Nearly all eight occupied an area in the southeast portion of the second floor. When we add the positions of those who were injured to those who died, we see that the farther people were from

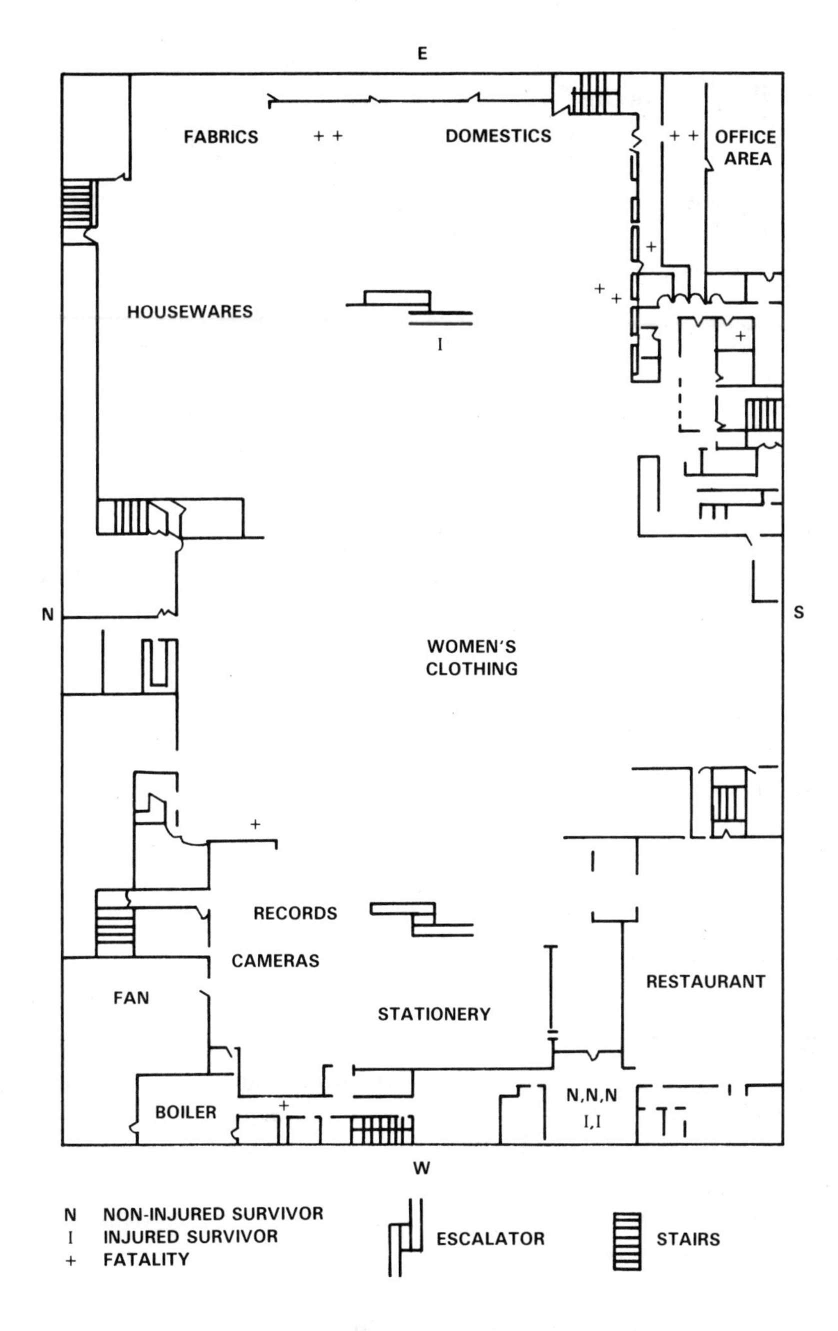


**Figure 4.1.** Second Floor of the Younkers Brothers Department Store.

the southeast quadrant of the second floor, the lower their likelihood of being injured or killed.

From the reports of the fire department, it appears that few firefighters suffered serious smoke inhalation injury. During their initial attempts to find the main body of fire, the heat and smoke pushed them back before they could even climb the escalator. They never came into contact with the early-generated, undiluted killer smoke. The firefighters complained that the smoke was unusually dense and acidic, even from a distance, and several commented on the number of airpacks they used: "I went out for my sixth canister before trying again, " one said.

The initial fuel that generated the killer smoke, and the origin area of the smoke, can be traced to the southeast corner of the second floor. This conclusion is based partly on the fact that fewer injuries and fatalities occurred as the distance increased between the person and the southeast corner, and that the density of the smoke prevented even firefighters equipped with self-contained breathing apparatus from coming close to this area. The positions of the two men who died away from that quadrant indicate that they had time to attempt to escape and to move some distance before they succumbed. Presence in that southeast corner apparently meant immediate incapacitation.

## THE INVESTIGATION

Shortly after the fire, the NFPA sent one of its investigators to West Des Moines to reconstruct the fire. The NFPA tries to issue a report for every fire of ten or more deaths. The report that resulted from this visit and from extensive cooperation with the Des Moines Fire Department described the store in great detail. It also included the occupancy of the store at the time of the fire, and a description of the event itself from the low-grade explosion in the plenum (the space between the ceiling of one story and the floor of the one above it) through the extensive damage that occurred after the merchandise and the furnishings were involved in the fire. The investigator, who was a fire science expert, was confronted with an explosion in the plenum; with an account of greasy, extremely dense and black acidic smoke bursting from the ceiling; and with most fatality victims showing evidence of nearly instantaneous incapacitation.

No theory of origin or initial fuel came out of the NFPA report. The Des Moines Fire Department, with fewer resources available to it, and less pretension, turned up more evidence and tried harder than the NFPA, whose report was one of the lamest in the fire reconstruction literature. One of their main conclusions was that the store wasn't sprinkled! Sprinklers in the habitable space would have had no effect on the events that occurred above the ceiling or on the generation of the killer smoke. The NFPA report was not only inadequate, but also unconvincing in its parroting of conventional platitudes meant to mislead.

The reasons for this inadequacy can be blamed on the fact that NFPA fire investigations have fallen behind the times. The investigators generally have inadequate qualifications for today's fires, which are heavily influenced by new materials and architectural/interior design techniques. The NFPA attitudes about responsibility for ignition, spread, and results of fires haven't changed with the times, either. Investigations are, thus, slanted toward "proving" that codes were violated and/or that people were careless. Other factors are simply ignored.

The Des Moines Fire Department and its consultant acted on the description of the dense, acidic smoke and on the placement of the explosion in the plenum. They generated a hypothesis that the freon in the air conditioning system interacted with the aluminum of the ducts and started a fire, and that this fire created hydrogen chloride. In pursuing this hypothesis, they found a number of relevant clues.

- An opened container of boiler cleaning fluid had a much lower pH (was more acidic) and higher chloride content than a container bought new from a store.

- The concrete in the store was extensively spalled. It was damaged by hot, acidic fumes and smoke. Heat alone could not have spalled the concrete. It showed the effects of being eaten away by acid.

- The security manager of the store complained that the merchandise wasn't actually burned by flames, but surface-burned by contact with the soot.

- The worst damage occurred in the southeast quadrant of the second floor, with secondary damage from fires igniting near the outlet of the air supply ducts to the habitable spaces.

The interviews with the survivors indicated that first there was a pressure wave of air (described by most as a strong wind) and then a cloud of thick black smoke that came from the southeast quadrant of the second floor. The people on the western side of the second floor saw the smoke coming from the east, and the people on the first floor saw it first issue from the escalator opening from the second floor. Several witnesses said that the store clocks stopped at about 9:30 a.m. The electrical system, by that time, had completely broken down, but the smoke and heat did not appear until, at the most, perhaps two to three minutes previously. One witness had glanced at a clock five minutes before the pressure wave and noted that the clock was running and that the time was 9:20 a.m.

The Fire Department also negated several other possible causes of the fire and fuels: the mineral tile ceiling and its mastic, the boiler and boiler cleaning fluid, and a natural gas leak. The department had the gas pipes tested for leaks and found nothing that could have resulted in such a fire. The investigators also had the other materials tested, with negative results. The possibility of arson was investigated. They also investigated a gas line leak into a nearby creek, wondering if the gasoline vapors could have permeated the air in the store. The Des Moines Fire Department listened to all witnesses, including people who had shopped at the store and thought they had smelled gas up to a week before the fire. (These odors seemed to be vehicle fumes from the parking lot.) All avenues were explored and a mound of data was amassed, including scores of photos of the store right after the fire. The complete report also contained all the autopsy protocols by the coroner's office.

Table 4.1 lists the ages of the victims, blood analysis results, and other observations made by the medical examiner's office. By correlating the age with the carboxyhemoglobin percent (the percent of the hemoglobin attached to carbon monoxide), we can see that the older a victim was, the lower the percent of carboxyhemoglobin. The conclusion to

**Table 4.1.** Results of Partial Autopsies.

| Sex | Age | Carboxy-hemoglobin* | Cyanide** | Observations |
|---|---|---|---|---|
| M | 42 | 80% | 0.3 | Soot in trachea; Methane in blood. |
| M | 67 | 88% | | Soot in trachea; Methane in blood. |
| M | 28 | 80% | | Soot in trachea. |
| M | 62 | 52% | 3 | Soot in trachea; Methane in blood. |
| F | 28 | 76% | 0.2 | Soot in trachea; Methane in blood. |
| F | 28 | 78% | | Soot in trachea; Methane in blood. |
| F | 51 | 69% | | Soot in trachea. |
| F | 69 | 48% | | Soot in trachea; Methane in blood. |
| F | 67 | 53% | | Soot in trachea; Methane in blood |
| F | 60 | 64% | | Soot in trachea. |

*Percentage of hemoglobin attached to carbon monoxide. Lethal level is generally considered to be 50 percent.

**Cyanide in bloodstream in units of micrograms/milliliter.

be drawn from this observation is that the victims lived a duration of time in inverse proportion to their ages. The older victims died first, while the younger victims were able to hang on a little longer and thus to breathe in more of the poisonous smoke. The carboxyhemoglobin percentage indicates how long the person kept breathing after incapacitation.

All of the victims had significant amounts of soot in their respiratory tracts as far down as the partial autopsy opened. The one victim that was later exhumed and given a full autopsy showed soot all the way into the deep lung. In addition, a forensic pathologist with much experience in this area determined that the tissue abutting the soot depositions was irritated and showed signs of incipient inflammation. The conclusion that can be drawn from this information is that the victims inhaled large quantities of irritating soot deep into their respiratory tracts. Indeed, the nostrils of most victims appeared to be plugged with soot.

Most of the victims were extensively burned because the fire spread from the plenum. The medical examiner and the forensic pathologist both concurred that the deaths occurred before the flames touched the bodies.

The Des Moines investigators did not pursue the possibility of latent injuries among the survivors. (Latent injuries are injuries that appear some time after a catastrophe, but then can be traced back to that catastrophe.) This line of inquiry was outside their expertise and probably never occurred to them. The only medical records that appear in the two volumes of the investigation report are those from the immediate treatments of lacerations and bruises.

The plaintiffs' counsel, on the other hand, pursued several lines of inquiry in addition to those researched by the Des Moines Fire Department. They analyzed the medical records of survivors, soot chemistry, the store's electrical usage records for that week, and damage analysis by fire reconstruction experts. The team of experts for plaintiff's counsel included chemists, physicians, toxicologists, electrical engineers, fire reconstructionists, and combustion chemical physicists. Meetings included discussions of specialized areas such as how to get a representative sample of soot particles, the dynamics of the pressure wave from the ignition, and what in the smoke and fumes would lead to nearly immediate incapacitation.

One of the early pieces of data assembled involved the soot. A clock and a lab coat taken out of the building within the first five minutes of the explosion had soot on them. Electron microscopic examination of the soot revealed that it did not resemble soot from wood, coal, oil, or natural fabric. Elemental analysis of this soot revealed that it contained bromine, chlorine, antimony, lead, and zinc. Out of the universe of possible elements, this is a very small set and can be traced back to one product present in the plenum—PVC. PVC contains brominated fire retardants, antimony oxide and lead stabilizers, and zinc organics as stabilizers and lubricants.

The injuries reported by a few of the survivors and documented by their medical records revealed no resemblance to the simple carbon monoxide poisoning that traditionally typified old-fashioned fires in natural fuels. Two young people, a man and a woman, suffered frequent respiratory infections of prolonged duration. The woman also felt tired all the time, had upper respiratory membrane swelling and reddening that included sinus troubles, and was troubled with lower back pain. The young man was suddenly afflicted with high blood pressure and two types of heart problems, tachycardia and a systolic ejection murmur.

An older man came down with chronic coughing, phlegm production, and winter bronchitis. This man was taken to the hospital hours after the fire because of delayed adult respiratory distress. He could not get through two sentences of his deposition without coughing. He was one of the few injured survivors who did not sue, although he gave testimony and showed that he later realized that his health deterioration was connected with his smoke exposure. By the time the realization came to him, the statute of limitations had passed.

One man who was hospitalized a couple of hours after the fire had a number of pre-existing conditions that the attending doctor discovered: leukemia, a brain tumor, and emphysema. The smoke exposure had nothing to do with these conditions. However, his initial symptoms drove him to seek hospitalization. These symptoms were a sudden onslaught of acute coughing and worsening shortness of breath and were a result of smoke exposure. This showed that his already weak health status was further compromised. It would be impossible to conclude whether or not the exposure in the fire had an effect on the length of this man's life. He had gone on from day to day, working and going home to his wife, until the fire. How long that routine could have lasted, had the fire not occurred, would be mere conjecture.

The reconstruction of the fire by a team of experts assembled by the plaintiff's counsel appears in the following paragraphs.

An unspecified electrical malfunction occurred in the southeast quadrant of the second floor plenum. Certain clues hint that this malfunction initially may have started in the new computer cash register system. In any case, the wiring involved in the malfunction overheated and decomposed. The other wiring in the area also overheated and decomposed directly from the heat radiated from the malfunction and indirectly from the hot gases generated. The overheating spread and continued for at least many hours and produced both corrosive and combustible gases.

This occurred before workers arrived Sunday morning. One of the first to arrive went to the boiler/utility area of the store and turned on the air circulation fans. This caused the oxygen level in the affected quadrant to rise until the oxygen/fuel ratio reached the explosive level. Then the flaming stage of the fire began. The explosion knocked down the ceiling tiles and freed the soot and decomposition gases that had accumulated during those hours of overheating. The pressure wave and the expanding gases moved the smoke and fumes through the store.

As the smoke and fumes moved through the store, they became diluted, and the acid reacted with the surrounding surfaces so that the smoke became less harmful. Those people present at the point where the smoke was first released were exposed to lethal concentrations of decomposition/combustion products almost immediately. Those farther away either became incapacitated and died later or were injured. Those farthest from the first release of the smoke were either injured or got away without permanent injury, depending on susceptibility and length of time exposed to the smoke. Thus, we could see a dose-and-response relationship.

The long decomposition period also explains the fire that was seen by the firefighters and attributed to natural gas. Natural gas does not burn with a green flame, such as the firefighters saw. Chlorine, however, imparts a green color to flame. A mix of chlorinated and non-chlorinated hydrocarbons, which arise from pyrolyzing PVC (see Table 1.1), would behave like a natural gas fire and impart a green flame.

The air handling system circulated the smoke and heat from the fire through the building because the plenum acted as a return air pathway. Secondary fires and smoke damage, thus, occurred at air vents, after the flame stage began and the victims had been incapacitated by the initial wall of smoke from the ceiling.

## THE LITIGATION

The defendants in the Younkers Brothers Department Store fire litigation were the manufacturers of the wiring that was present in the plenum and the manufacturers of PVC resin, which goes into electric wire insulation. These are some of the largest companies in the world (Exxon, Tenneco, Occidental, Diamond, Shamrock, Rheinhold, and AT&T). In fact, PVC accounts for 1 percent of the United States gross national product. To make an understatement, these companies had the financial and technical resources to defend themselves and to make a case that would have shifted the burden onto some other product, if such a case could be made at all.

Instead, these enormous companies either settled or bumbled in trying to make an alternative case to explain the events in this fire. By the time of the trial, the majority of the original defendants had settled and had contributed various amounts of money to the plaintiffs. The plaintiffs included not only the estates of the fatality victims and most of those injured, but the department store and its financial institutions.

The remaining defendants tried to offer the following explanation of the origin and spread of this fire: the freon from the air conditioning refrigeration unit leaked out and interacted with the boiler fluid and the aluminum ducts and louvers. This reaction released both hydrogen chloride and heat. The heat set the lint in the air ducts and louvers on fire, spreading the fire. Therefore, the fire had to have started in the opposite side of the store from the death quadrant, the northwest corner where the boiler room was.

This scenario lacked credence because of all the eyewitness accounts that placed the main fire activity in the southeast corner of the second floor, and because the reaction of the freon would also have released hydrogen fluoride, which etches glass. No damage from glass etching was seen after this fire.

Another tactic that raised credibility problems for the defendants is that in their attorneys' opening statement, their lead counsel asserted that PVC was so nontoxic that it could be eaten. The jury simply didn't swallow it! This attorney later published a paper complaining that the personalities of expert witnesses determine jury reactions, not the technical arguments of these witnesses. In this case, this man had neither the technical arguments nor the tactical sense of how to get a jury to listen to what he said. Juries are notorious for judging credibility. Bless them for their sense of truth.

The eight-person jury found, in a seven-to-one verdict, that the Younkers Brothers Department Store Fire was an electrical fire that initially decomposed and burned the polyvinyl chloride wire insulation of the electrical systems present in the second floor plenum. It further found that the deaths, injuries, and damage resulted from the combination of the combustibility and toxicity of the PVC wire insulation. The one juror who did not agree with the majority later explained in a newspaper interview that he was really looking for proof beyond a reasonable doubt, not a conclusion based on the preponderance of the evidence. What he did not seem to understand is that the former criterion applied to a criminal verdict, not a civil; the latter applies to a civil verdict. Thus, the one dissent represents a misunderstanding of how evidence applies in a civil trial.

A number of corporations had reached settlement agreements before the trial. After the initial verdict, nearly all of the defendants chose to settle rather than stand trial for assignment of damages. In fact, only one defendant held out for a damage trial and a series of legal maneuvers, including attempts to change courts from state to federal level. This corporation (Goodrich), by the way, had as its counsel the same man who told the jury that PVC is okay to eat. This lawyer fits in well with the PVC industry, which seems dedicated to ignoring both truth and social responsibility.

The pre-and post-trial settlements hold this meaning for the public: these vast corporations, with all their financial and technical resources, knew that they could not pull together a credible case for their product. They also knew that the story they were telling of the origin and spread of the fire resembled Swiss cheese. Lawyers from all over the country had been hired by these bloated companies, and they still couldn't make a credible case against a small number of local attorneys representing plaintiffs and the plaintiffs' experts. This nose for truth remains one of the miracles of the citizen-juror and a foundation of whatever justice we have in our society.

## THE RESULTS

The Younkers Brothers Department Store Fire offers us lessons both with respect to the behavior of materials in fires and with respect to public policy on that behavior. We can also learn about the interaction between these technical and policy factors.

Generally, codes were not violated in the Younkers Brothers Department Store. Exits were not blocked. The store was not overcrowded. The wiring had been installed according to the Life Safety Code (a code created by the NFPA), except in one particular way—the plenum was the return air pathway, and combustible materials are not supposed to be installed in the return air pathway to prevent the spread of smoke, fumes, and heat around the building via the air ducts. No code provision that is intended to prevent fire occurrence was violated, yet the fire ignited. The fire ignited because of the nature of the material and because of the nature of electrical systems. This material (plasticized PVC) is so unstable that even in the absence of code violations, it will decompose and produce combustible gases. Our codes are geared to materials of the stability of wood, paper, wool, and rubber. The codes that are sufficient for these older materials are not sufficient for synthetic polymers—especially for PVC.

A material that has the instability of PVC invites disaster when it is used in large quantities in hidden spaces such as walls and ceilings. Its use in hidden spaces in conjunction with an energy source only compounds the hazard. The early signs that the material is decomposing are invisible. Heat buildup can't be felt. The process of decomposition can continue for a long time and reach lethal stages before anyone is aware of anything unusual at all. Furthermore, the interactions of the different building systems (such as heating, cooling, and computer systems) and their effect on the decomposition process can't be detected until after the disaster. Ordinary actions, such as turning on the ventilation fans, could trigger sudden changes because of the undetected decomposition process.

The Younkers Brothers fire litigation is one of a series of lawsuits that exposes the paradox of a wealthy industry with all of its power and big-time lawyers, not being able to make a credible case for its product. And the fundamental reason why a credible case cannot be made is that the product is dangerous and defective. It cheats consumers because it degrades and loses its initial physical properties, and it endangers public health and safety because of its chemical instability. If this multibillion dollar industry could have made a case, and if the product were at all defensible, then the whole dynamic of the litigation process would have been entirely different. The industry would have been able to present data that were not ambiguous in a simple way without all the convolutions and recourse to buzzwords such as "tradeoff" and to misleading statements such as, "Everything that burns is toxic." The industry would have been able to dig into the facts of the fire and arrive at a scenario that would have been very difficult to attack. But this could not happen, because the fundamentally dangerous nature of PVC and the wealth of evidence precluded coming to any other conclusion than that of the jury.

The industry persuaded the court to grant a media gag. A media gag prevents the reporting of the details and outcome of a trial and states that none of the parties involved can speak to the press about the litigation. Through such devices as media gags and sealing of court records in event of settlements, the industry conceals what should be available to the public: the record of the litigation and the events leading up to and including the fatalities, injuries, and damage. In the case of defective and dangerous products, these legal devices are immoral and possibly fundamentally illegal, because they conceal illegal threats to public health and safety, such as deliberate risk of exposure to lethal levels of hydrogen chloride. But with legal devices such as gag orders, corporations can make public statements about how no death was ever connected with their product, all the while knowing that the deaths have been recorded in the sealed records and gagged suits. In this way, the producers of killer products evade the eye of the Federal Trade Commission, the Consumer Product Safety Commission, the Center for Disease Control of the Public Health Service, the anti-racketeering projects of the FBI, and the Environmental Protection Agency Bureau of Toxic Substances.

The inability to gather the data in the sealed records and gagged suits prevents development of desperately needed public policy. This policy could save others from becoming similar fatality victims and injured survivors. The failure to see the causes of these deaths and injuries means that the public agencies fail in their mission to protect the public and to educate the public to protect themselves. Many injured survivors never connect insidious latent injuries with their inhalation of the strange smoke. They neglect to tell their doctors that they had a smoke inhalation episode that involved fuming or burning plastics. The injury is attributed to something else, or labeled idiosyncratic. With respect to the gags and seals, the lesson is that they form an insidious latent injury on the body of justice and public health and welfare in this country, and the circumstances for applying them should be severely limited by law.

The Younkers Brothers fire also illustrates the strain that plastics fires place on relatively small fire departments. At the height of the fire, only one pumper remained available in case there was another fire in Des Moines. Two suburbs sent their units. A vast area was stripped of fire protection for hours. The lesson is, stated simply, that because of the nature of the fire, the firefighters were not only unable to reach the source and therefore control the fire, but were also unavailable should another emergency arise.

The most important lesson of this fire teaches us how quickly PVC fumes and smoke can lead to death, especially if the heating process has been hidden for hours. This unstable material is fundamentally, irredeemably dangerous. Just being in the wrong place at the wrong time dooms scores of people every year, as it did the unlucky people who were in or near the southeast quadrant of the second floor on that Sunday morning.

*CHAPTER 5*

# The Fort Worth Ramada Inn Fire

It was renovation time at the Fort Worth Ramada Inn. Recarpeting was part of the renovation. In Building B of that inn, old carpeting still lay on the floors, but new carpeting and padding were piled inside the doorway of the southwest exit of the first floor, ready to be installed.

Building B had two floors and contained about ninety guest rooms. Both floors had the standard long corridor with rooms on each side. The only exits were located at the ends of the corridors. Figure 5.1 shows the layout of Building B and the location of the carpeting and padding.

In the early morning hours of June 14, 1983, someone at the Fort Worth Ramada Inn was careless. Someone flicked a match or cigarette right onto the plastic-covered carpet and padding. The pile smoldered, decomposing the wax-like polyethylene plastic that was bagging each roll of new carpeting and padding. The fibers of the carpet and padding acted like wicks. The fire quietly smoldered, then began to burn like a gigantic wick.

The fire was discovered by a guard, who immediately used a fire extinguisher. A guest joined him, bringing an-

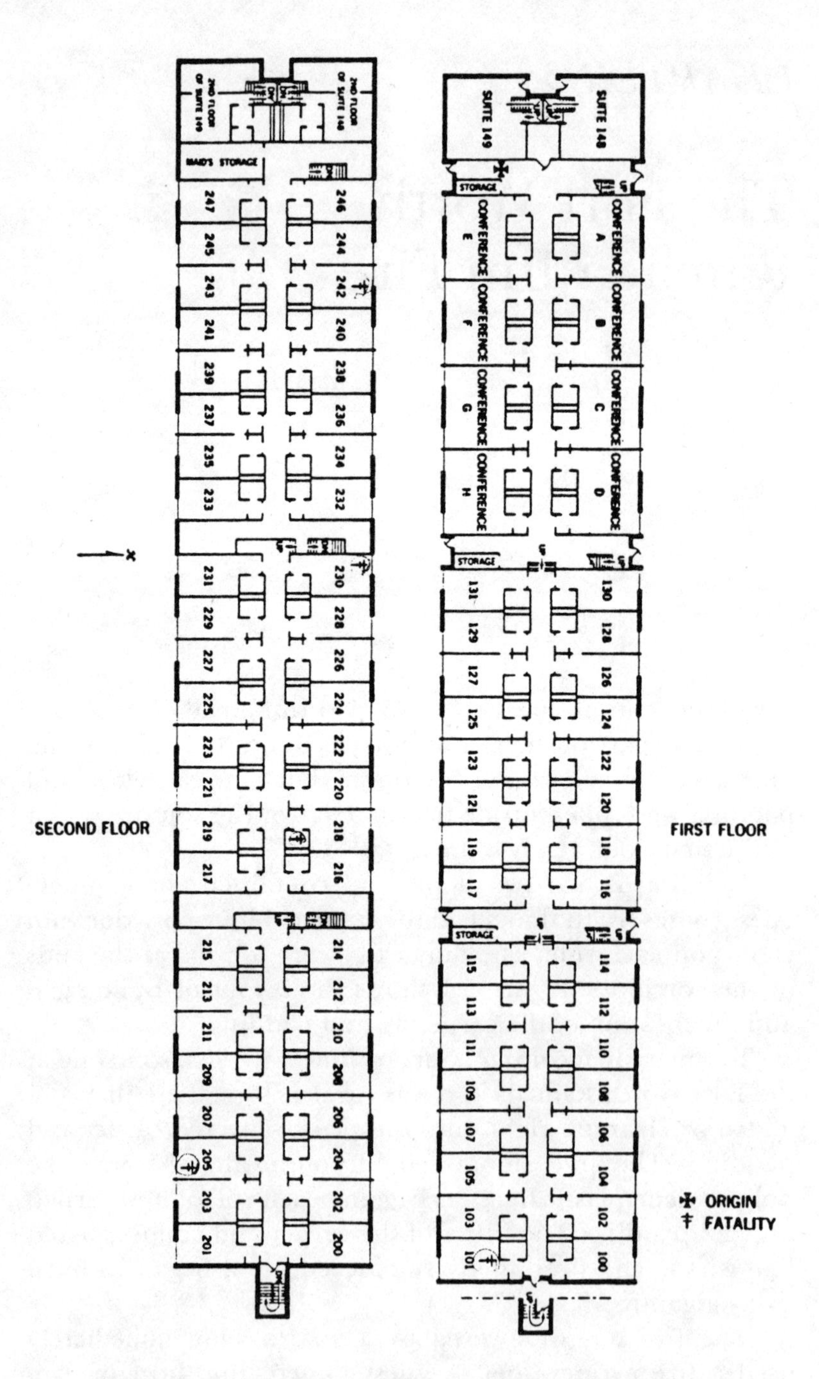


**Figure 5.1.** Floor Plan of the Fort Worth Ramada Inn.
Taken from NFPA report.

other extinguisher. The guard ran and informed the night staff of the fire while the guest held the fire at bay with the extinguisher. But the extinguisher gave out before the fire died. After all, what can an extinguisher do against a gigantic multi-wicked candle, like a 1,000 gallon fuel tank?

The fire spread to the rolls of carpeting and then to the installed carpet. It spread up the stairs, burning the wallcovering, and running horizontally along both floors. The guests were warned of the fire by an employee who drove around the parking lot, honking the horn of his vehicle. Guests had to jump out of windows—the corridors outside their rooms were filled with fire and smoke. Five guests never got out and died in their rooms. Thirty-six people were injured.

The NFPA sent an investigator to the scene who duly noted all violations of the existing code and the most likely scenario of ignition and spread. Part of the account below is from the NFPA report. The Fort Worth Fire Department also investigated the fire and added its own thoughts about the factors responsible for the injuries, deaths, damage, and rapidity of events.

## THE FIRE

The fire started in the pile of new carpeting and padding. The position of these materials is shown in Figure 5.1, indicated with a circle. This location is strategic: it is at the bottom of a stairs (chimney effect), at the end of a long corridor (corridor effect), and at one of the few exits to the building. All fire analysts involved in the investigation and litigation agreed that combustible materials should never be placed at an exit. Few, however, commented on the multitude of sins that were committed when the materials were placed here. NFPA noted the chimney effect, but the corridor fire fad had already faded by the time of the Stouffer's Inn Fire of 1980 and the technical staff no longer saw every fire in a corridor as a corridor fire even when it was one. (In 1975–1978, it was a short-lived fad to call a fire a corridor fire.

After that, the phrase "corridor fire" disappeared from NFPA reports. The existence of these fads in the investigation of fires points to the inadequate technical staff of the NFPA.)

All analysts agree that the fire spread very rapidly. The Fort Worth Fire Department noted the nature and composition of the carpeting and padding as possible factors, along with the vinyl wallcovering along the corridor and stairwell. The wallcovering showed a burn pattern that indicated heavy decomposition and charring during the early part of the fire, when it was confined to the piles of material. The smoke preceded the flames both up the stairs and laterally along the two corridors of the first and second floors. The flames first rose up the wood stairs, pulled along up the rising, hot, combustible gases. Damage patterns showed that the fire raged, especially on the second floor. Part of the cause of the difference in the amount of damage on the floors is the chimney effect and the trapping of the heat and gases under the second floor corridor ceiling.

The Fort Worth Fire Department and NFPA investigated the possibility of arson by both following up leads that disgruntled former employees were in the vicinity and analyzing samples for accelerants. Neither avenue of investigation revealed even a shred of possible arson evidence. After studying the burn pattern, one of the experts retained in the litigation by the motel asserted that arson had occurred. He had seen certain places in the corridor with heavier burns than others, and believed that these places showed where fires had been set with an accelerant. This was very thin evidence at best. The deeper burns could also be explained by later burning and dripping down of ABS (acrilonitrile-butadiene-styrene plastic) pipe from the ceiling above and by renewed air supply from broken windows.

The fire spread along the corridors of both the first and second floors, but did the most damage on the second floor. It spread into some of the rooms because the doors offered little protection. Because there were gaps between the floor and the door and the frame and the door, the door itself could not protect the room. By the time flames actually threatened rooms and their occupants, the occupants had either escaped through the windows or died from smoke inhalation.

At the time of incapacitation leading to death, the major fuels were carpeting, carpet padding, and wallcovering. Both installed and soon-to-be-installed carpeting eventually burned, but the early smoke came solely from the piled-up rolls, installed carpeting in the vicinity of these rolls, and the wallcovering along the stairs and in the adjacent corridor.

During the time of firespread, one of the hotel personnel tried to notify guests of the danger by driving around Building B, honking his horn—there were no fire alarms in the hotel. The switchboard operator called rooms until the power died. The Fire Department received the alarm at 3 a.m. and dispatched the first alarm one minute later. By the time the fire unit arrived, guests were already jumping out of the windows.

The guests broke the fixed-pane windows with various pieces of furniture, usually chairs. Some, in sheer panic, immediately jumped out and received lacerations from shards of the glass that they themselves had just broken. Others threw mattresses out first, both to cushion their fall and to protect against cuts. Still others put on shoes and wrapped themselves in blankets or sheets as protection. In any case, many received lacerations, sprains, and broken bones. The emotional force behind these immediate leaps didn't come merely from hearing the horn-blowing and the rising sound of the sirens. Smoke had penetrated the rooms—a choking, poisonous, dense smoke that carried with it the message "Leap and live or stay and die."

Firefighters arrived within two to three minutes of the dispatch and saw smoke and fire at the east end of the building. A second alarm was transmitted and dispatched immediately. As more firefighters arrived and as they could turn their attention to fire control as opposed to guest rescue, they realized that they had mistakenly assumed that the east end of the building was the main fire location. After spending time at the east end, they discovered the main body of flame at the west end. The blaze eventually received a total of five alarm assignments. A triage center had to be established for the injured, and the ambulances were overwhelmed by the sudden demand for service.

The following quotes are from witnesses and investigators. They help to reveal the tenor of events the night of the fire.

> This investigator arrived on the scene at approx. 3:40 (a.m.) traveling south on Beach from I-30. I observed moderate smoke crossing I-30 west of Beach Street.... Heavy black smoke was pouring from the upper windows in several of the rooms on the south side of the complex. No flames were visible at this time. There were several people wandering around with various cuts and injuries and firefighters on the scene were assisting them to the triage area east of the Ramada Inn.

> The companies responding to the first alarm were p-14, Q-20, A-20, L-14, D-4, and M-6. They found heavy smoke coming from several windows on the second floor and began setting ladders on the south side of the building to aid in the rescue of the hotel guests that were attempting to jump from the second floor windows. Second alarm companies, P-5, A-5, P-1, L-1, U-1, S-2, S-201, P-2, and C-2. The third alarm companies that responded were P-19, P-10, L-2, D-1, P-3. The fourth alarm companies responding were P-21, and P-17, D-5, and the fifth alarm brought out P-8, P-4, P- 12.

> Initially there was a great amount of confusion with the heavy black smoke pouring from every open window, especially from the east end of the fire building that led first arriving firefighters to suspect that the fire was at the east end of the building. Occupants of the various rooms were throwing chairs and tables out the plate glass windows in an attempt to escape the fire, heat, and smoke that was filling the halls on both the ground floor and the second floor of complex "B."

> The fire appeared to have traveled from the center of the entrance hall on the first floor, west end of the complex "B" and spread upwards by way of the stairway on the northeast wall of the west hallway and spread rapidly east down the long unobstructed hallway. No guests interviewed were able to escape using the hallway due to heavy black smoke and heat that was present when they awoke during the fire. The fire was aided in its extension by the existing carpet on the floors, the rubber padding beneath the carpet, the vinyl covering on the walls, and the vinyl-covered doors and door facings. In several rooms, the padding and carpeting had burned four to six inches under the door. The building is approximately 60 feet wide and 300 feet long with the hallways running the length of the building with no fire doors separating the different sections on either floor. The long hallways apparently acted as a chimney drawing the smoke, heat, and fire throughout the building." These are from reports by R.L. Frazier, fire marshal of the Fort Worth Fire Department.

The following is from the deposition of retired Deputy Chief Jerry Smith of Oklahoma City, who was a guest at the inn during the fire.

> When I first got up, my first recollection that there was a problem was I smelled smoke, and then I glanced toward the ceiling and I could see the black smoke on the ceiling. . . . I could see the thick black smoke rolling in from the top of the door, puffing around the sides and coming in the bottom. And it was pretty black and real thick. . . . Well, definitely it burned my eyes, and would cause you to cough very quickly. . . . With the way that smoke was rolling in the room, or seeping in the room, I think, and again this is an opinion, we needed to exit that room within two or three minutes. Because the toxic fumes entering the room would have been—could have very easily overcome us within that time frame.

## THE VICTIMS

Five people died in the fire, four men and one woman. Figure 5.1 shows the location of the fatalities: four on the second floor and one on the ground floor. Their ages ranged from twenty-five to fifty-four.

The medical examiner conducted full autopsies on the fatality victims and made many important observations. The respiratory tracts of all the victims were loaded with soot. Soot had even penetrated to the pleural side of the lung in some of the victims, which is the side of the lung toward the inside of the body rather than the side open to the air. All lungs showed severe reactions: frothy edema (liquid), hemorrhaging, destruction of tissue, and congestion with blood. Those with soot on the pleural side also had hemorrhaging through the lung to the pleural side. The upper respiratory tract also contained soot and showed reddening and edema of the larynx, signs of irritation.

Most of the organs were also observed. Some showed signs of irritation and of the type of blood congestion that is indicative of heart fibrillation (which is when part of the muscle moves or twitches, but the entire muscle is unable to function). These signs were that the whites of the eyes were "injected" (reddened), many internal organs were congested (engorged with blood), and brain edema. Brain edema shows that the heart not only pumped a great deal of blood to the brain, but also sent a special diuretic hormone to the brain that modified the blood/brain barrier and allowed blood fluids to enter outer portions of the brain. Brain edema was also evident in the victims of the Stouffer's Inn fire (Chapter 7), the MGM Grand Hotel fire (Chapter 6), and a high-rise fire in Toronto in 1983. Brain edema hints that chemicals in the smoke can get into the brain because of the breach in the blood/brain barrier. Remember the evidence of neurotoxicity in survivors of the other fires in this book (see Chapter 3 and Chapter 6). The heart itself may open the brain to smoke poisoning with its barrier-breaking hormone.

The medical examiner noted that all the individuals showed physical traits appropriate to their ages. All were well-nourished, well-muscled, and free of signs of a history of unusual stresses. The middle-aged men had signs that are typical of aging in our culture: incipient atherosclerosis, gall bladder problems, and general flabbiness.

Let's make a brief digression here about age and physical condition. Not long ago, a study was reported in the magazine *Natural History* about a South American tribe that was living with few modern amenities. They took few sanitary precautions. The tribe members had many parasites. Nearly every organ had a parasite living in it. The authors of the study reported that the people they examined showed remarkably good health for such a parasite load. They failed to understand the significance of the fact that no one lived beyond age thirty.

We live long. We begin to age at twenty-five years, but the average length of life is around seventy-five. There are some combustion toxicologists employed by the plastics industry, such as Ted Radford of the University of Pittsburgh, who maintain that because more atherosclerotic people die from lower doses of carbon monoxide than people with youthful blood vessels, the manufacturers of these fuels should be let off the hook—it is the people who are flawed, not the product.

On the contrary, the fact that a large sector of our population is especially vulnerable to carbon monoxide places greater responsibility on product manufacturers. Irritants that incapacitate people and hold them powerless in a fire-generated atmosphere must be carefully avoided, as must the substitution of traditionally noncombustible materials with organic materials. Companies doing intensive market research into changes in population age structure for selling purposes literally commit a crime when they neglect the safety implications of those changes. Digression over—let us return to the Ramada Inn.

Blood chemistry tests done on the fatality victims showed that none had been incapacitated by either alcohol or drugs.

Table 5.1 shows the levels of carbon monoxide and cyanide found by standard analytical methods. This table also shows the ages of the victims. In fires, the concentrations of carbon monoxide in the fatality victims are often negatively correlated with their ages. This was true in the Stouffer's Inn and the Younkers Brothers Department Store fires. However, note that one of the youngest victims of this fire had low levels of both carbon monoxide and cyanide. This man was in a room that was very close to the fire origin. The levels of carbon monoxide and cyanide seem patterned according to distance from the origin more than the age of the victim. The two victims farthest from the origin of the fire had no detectable cyanide and the highest levels of carbon monoxide. We shall comment on these apparent patterns when we interpret the implications of the data.

Although many of the injured sustained lacerations and bone breaks, because all of the survivors jumped out of windows, the most frequent injury was smoke inhalation. That is all the more remarkable in light of the fact that of all the fires in my data base, the Fort Worth Ramada Inn blaze is the only one during which all escaped through windows. Even in the horrendously fast Stouffer's Inn fire, some of the survivors raced the fire down corridors and exited through doors.

The experts working on behalf of the plaintiffs obtained several of the hospital records on the injured. The most consistently reported problems were abnormal blood gases and blood pH, breathing difficulties, headache, blood pres-

**Table 5.1.** Blood Chemistry of Fatality Victims.

| Sex of Victim | Age of Victim | Percent of Carbon Monoxide (CO) in Blood | Cyanide* (mcg/ml) |
|---|---|---|---|
| M | 26 | 28% | 2.5 |
| M | 25 | 77% | 0 |
| M | 54 | 62% | 3.0 |
| M | 46 | 75% | 4.5 |
| F | 27 | 79% | 0 |

*Cyanide measured in microgram/milliliter.

sure instability, and heartbeat irregularity. All of these are symptoms described by the injured survivors of similar previous fires. As with the Stouffer's Inn fire, all survivors were out of the building within a few minutes after the fire was discovered. Even so, thirty-six people required medical attention, mainly for the effects of smoke inhalation. The toxic smoke killed those who couldn't escape immediately and injured those subjected to extremely brief exposures.

Let us digress a short second time to consider irritants. Toxicologically, an **irritant** is any chemical that triggers and sustains inflammatory reaction, such as reddening, expansion of blood vessels, a burning feeling, production of protective fluids such as tears and mucus, and sometimes leaking of blood vessels and mobilization of certain types of white blood cells. **Corrosive** irritants also attack and kill tissue, and cause hemorrhaging along the respiratory tract. Citizens generally don't understand the importance of irritants as degraders of health. One reason for this gap has to do with a special property of irritants—their time/dose relationship is different from that of other toxic chemicals. A long exposure to a low dose of irritant is not the equivalent of a short exposure to a high dose, as it is for other kinds of chemicals. High doses of irritants are much more damaging than long, low doses, and cause a much wider spectrum of health effects than low doses. Single high doses can lead to permanent injuries or even death, although the dose may be quite brief.

Table 5.2 lists the ages, sexes, blood gases, blood pH's, and examination observations for those injured survivors whose hospital records became available to the plaintiffs' experts. As the reader can see, the blood pH's range from grossly acidic to decidedly basic (alkaline) with very few actually normal. Possibly those who suffered a longer, higher dose ended up with the acidic pH. It is likely that the smoke differed qualitatively from one area to another, as the differences in the blood chemistry of the fatality victims also indicates. High concentrations of reactive acidic gases remained only very near the fire origin during the short time people still occupied the building. Lesser concentrations were

**Table 5.2.** Examination Results of Seven Injured Survivors.

| Sex | Age | Concentration of Oxygen in Blood | Concentration of Carbon Dioxide in Blood | pH | Carboxy-hemoglobin |
|---|---|---|---|---|---|
| M | 58 | Low | High | 7.26 (L) | 14.4% |
| F* | 11 | Low | Normal | 7.41 | 3.5% |
| F | 47 | Low | Low | 7.29 (L) | |
| M | 58 | Normal | Low | 7.43 | |
| M | 58 | Normal | Low | 7.48 (H) | |
| M | 36 | Low | Low | 7.45 (H) | |
| F | 35 | Low | Normal | 7.39 | |

*Asthmatic attack
L = lower than normal
H = higher than normal

breathed by occupants farther from the origin and could be overcompensated for by respiratory and kidney mechanisms that balance pH. The acid breathed in by people near the origin of the fire may have been too large a quantity for compensation by the usual homeostatic processes of the body.

**Homeostasis** literally means "remaining the same." The body maintains balances of acid/base, oxygen and $CO^2/CO^3$ (carbon dioxide/carbonate ion), blood sugar, protein, other chemicals, and water. All of these components are interrelated and depend on many vital organs for proper balance: kidney, lung, heart, liver, spleen, bone marrow, and even the digestive tract. The three-part system of signals (1) the blood, (2) the autonomic nervous system, and (3) the neuroendocrine glands, has provided grist for the mill of physiological research for decades. Each generation of physiologists wonders anew at the intricacy, strength, and delicacy of this system. It can work wonders to keep us alive and well. It can also kill us by way of internal reactions that amplify the original pollutant that invaded the body. The industrial view that humans can take almost limitless chemical beatings is completely wrong. The homeostatic system is delicate, intricate, and largely unpredictable.

The other smoke-related injuries found in the Ramada Inn survivors, besides abnormal blood gases and blood pH, included high blood pressure, shortness of breath, heartbeat irregularity, and acute asthmatic reaction. The only person who showed no abnormality or large changes in blood pH among those whose records became available to the team of experts had an acute asthmatic reaction that, essentially, kept her from inhaling the contaminated atmosphere. What a preventative measure! She nearly died.

The other injuries listed in Table 5.2 related to smoke inhalation agree with those seen in the previous fires with similar fuels. Although respiratory injury and changes in blood gases and blood pH seemed relatively independent of age, sex, and other factors, cardiovascular reactions such as heartbeat irregularity and blood pressure upswing appeared mostly in people over age forty, and especially those over age fifty.

Unlike some of the injured survivors of previous fires, this group was not followed over time to see whether their injuries were permanent, temporary, or insidious and progressive. We can't form conclusions from the hospital admissions records alone. But we can make prognoses based on the similarities of the injuries listed in the records to the injuries of survivors of fires of similar fuels. We will say that you can bet some of the injured survivors of the Ramada Inn Fire sustained permanent, insidious, and progressive injuries from the fumes and smoke.

## THE CAUSE

The heat- and fire-triggered behavior of the polymers in the hotel led to much of the outcome of the Ramada Inn Fire. Polyethylene or polypropylene wrappings around the carpeting and padding rolls formed an almost perfect initial fuel. Closely related to candle wax, polyethylene melts and vaporizes readily, especially in the presence of a "wick." The carpeting and padding fibers then began to act like wicks and became permeated with liquid and gaseous polyolefin. (Polyolefin becomes liquid when heated to melting temperature and gaseous when heated to vaporization temperature. Vaporization temperature is below ignition temperature.) The aerodynamics typical of fires then became established, carrying the heat and the products of pyrolysis and combustion up to the second floor via the stairwell. The vinyl wallcovering along the stairwell, exposed to growing heat and the chemicals of pyrolysis/combustion, began to emit hydrogen chloride and plasticizer, a highly combustible organic compound. A cloud of thermal products began to move laterally along the ground floor and second floor corridors. This cloud contained carbon monoxide, hydrogen cyanide and nitrogen oxides (from carpeting), hydrogen chloride, plasticizer, and possibly styrene from the padding and miscellaneous hydrocarbons. At this early stage, the cyanide and irritants are present at high concentrations. As the cloud passes down the corridors, these constituents decrease in concentration because of dilution and reaction with surfaces.

In fires fueled by plastics, the combination of the elements that are emitted could be poisonous, like the fumes produced by the combination of ammonia and chlorine bleach. Both of these products carry a label that warns against mixing the two. Fires that have fuels of a mixture of plastics (such as the Ramada Inn fire) may expose the building occupants and emergency service workers to the smoke equivalent of mixing ammonia and chlorine bleach. Mixed fuel loads typically include plastics based on nitrogen, some based on chlorine, some with bromine and/or chlorine fire retardants, some with phosphorus-based fire retardants, and some based on sulfur. Cadmium, lead, zinc, antimony, and silica commonly stabilize, color, or reinforce the fundamental carbon skeletons as additives.

Combustible materials must pass several tests, including flammability and fire spread. These tests address the materials one by one. We have no idea whether the products of decomposition or combustion of one material may influence the flammability or fire spread behavior of another because they are only tested individually, not in combination. In particular, we have no idea whether the products of different materials interact to alter the toxicity of the smoke in unpredictable and unsafe ways. Because of the example of ammonia and chlorine bleach, we should assume that this interaction can and does occur in smoke. The pile of carpet and padding rolls couldn't have been placed in a worse position for smoke and fire spread. The stairwell acted as a chimney, and the halls on both floors made the fire a textbook corridor fire with all the stereotyped attributes: rapid smoke spread, rapid heat spread, and rapid fire spread.

Another way the design interacted with the material involved the dilution of the smoke with distance from the origin, and the reaction of the acids in the smoke with the long surface areas. The first fatality with respect to distance showed a nonlethal level of carbon monoxide, a moderate and possibly lethal level of cyanide, and severe tissue damage and edema from corrosive irritants. The victim must have been killed very quickly by the early smoke's unreacted, undiluted acid. (**Unreacted** means chemically free—the acid hasn't been diluted by any reactions with other chemicals). The two mid-distance fatalities showed lethal levels of carbon monoxide, lethal levels of cyanide, and tissue damage and edema. They must have lived longer to have accumulated a lethal carbon monoxide and cyanide level, and must have breathed more dilute and reacted smoke. (**Reacted** smoke is less free chemically—some neutralization has occurred by reactions with other chemicals.) The two fatalities farthest from the smoke must have breathed relatively diluted and reacted smoke. They had the highest carbon monoxide levels and no detectable cyanide, but showed tissue damage and lung edema typical of corrosive irritants. The concentration of irritants in the smoke by that distance could incapacitate rapidly and damage tissue, but could not kill quickly. The hall design prevented enough dilution of the smoke and slowing of smoke spread to have allowed escape without incapacitation.

As we said before, the materials and design of the building gave occupants only about 3–5 minutes to escape if they were to live.

## THE LESSONS

Now that we have examined the fire, its course, and its human losses (both deaths and injuries), what lessons can be drawn from it?

Many important lessons have been learned from fires. Government reforms have been made as a result of public pressure either at the ballot box or by other more direct forms of outcry. The combination of the general deterioration of goods and safety codes, resulting from corruption inside industry, and the awareness of the deaths and injuries caused by disastrous fires, triggered these citizen actions. High fire incidence and individual large fires formed an important component of citizens' cause for indignation. One fire that occurred early in the twentieth century was the Triangle Shirtwaist fire, which happened in New York City on March 25, 1911. In this disaster, 145 garment workers were trapped and killed. As a result of this fire, the first standards for fire exits were set, and other reforms in working conditions in the garment industry were made.

The reform governments took certain strong actions after attaining power. These actions included the upgrading of the building and fire prevention codes, standards for delivery of municipal services, and supervision of governmental employees in corruption-prone jobs, such as police officers and building inspectors. The upgrading of the codes and of the municipal services often followed advice derived from lessons learned in individual fires and from fire incidence patterns.

The present tactic of the NFPA and other governmental agencies is to ignore important contributions to fire ignition, spread, and deaths and injuries. This means that much-needed upgrading of codes, code enforcement, and emergency services will not move forward. The new hazards will place large numbers of people at risk, kill many people, and cause many serious, permanent injuries. The NFPA reports for most of the fires in this book, including the Ramada Inn Fire, maintain a deep and pointed silence about the role of the synthetic fuel load. The National Bureau of Standards, in its reports to the Consumer Product Safety Commission, (CPSC), uses its outmoded and inadequate toxicity test to conclude that nearly all synthetics are no more toxic than most other materials. The CPSC accepts these conclusions. This commission strictly maintains its jurisdiction over product safety and quashes attempts by state and local governments to ban dangerous products.

One unique lesson that can be learned from the Ramada Inn Fire is the importance of the safe storage of plastics. Outside of intended use, little fire safety attention is paid to the different aspects of the plastic life cycle. Yet when a product is stored there is usually a much higher density of material per unit area and per unit volume than there is in the intended use of the product. Some industrial insurance providers worry about warehouse storage of plastics and the need for extra-dense sprinklers. Little mention has been made by any fire safety agency about the on-site, non-warehouse storage of plastic materials. Placement of high volumes of high-energy synthetics into a small space raises the risk of a small fire that could grow very rapidly into a giant blaze, producing large volumes of choking, blinding fumes and smoke that arises suddenly.

It may be that the flammability and fire-spread tests used on these carpets have paved the way to their careless storage. The contractors and the building manager of the Ramada Inn had been assured by the manufacturers that the carpeting and the padding passed all these fire safety tests. When given such blanket and authoritative assurance, people often act as if the material were non-combustible, instead of unstable and high-energy. The contractors and managers probably didn't even consider the packaging material, the wax-like polyolefin sheeting. "The carpet and padding pass all the tests, so let's get this installation done pronto."

A corollary to this lesson involves the misunderstanding of the flammability and fire spread tests. These tests have severe limitations. They are unable to predict both fire behavior in a variety of real fire conditions and interactions between mixes of combustible materials in a fire. They give a false assurance of nonignitability and nonspread to those who don't understand just what the purposes of the tests are. When these tests are misunderstood, the result may be that materials are stored improperly. The plastics stored unsafely in the corridor exit at the Ramada Inn are an example of how dangerous improper storage can be.

Standard flammability and fire spread tests are needed to characterize materials in a general way. Other tests may also be necessary for understanding the behavior of the material in storage and in use.

A second lesson, seen clearly in this fire, teaches us that most smoke inhalation deaths and injuries occur very early in a fire from the early, irritant-laden smoke. The early smoke will incapacitate occupants very quickly and, if they are near the fire origin, even kill them very quickly, before they can accumulate lethal levels of carbon monoxide in their respiratory systems. Irritating early smoke from plastics-fueled fires is very dangerous.

A corollary to the second lesson is that the products that incapacitate and/or kill are the initial fuels. Not all the materials that are eventually consumed by the fire necessarily contribute to the deaths or injuries, especially in a fire such as the one at the Ramada Inn, during which all survivors left the building within three to five minutes of the fire's discovery. Easily decomposed plastics that emit high concentrations of irritants pose the greatest risk to building occupants. One of the worst culprits is polyvinyl chloride (PVC).

Lessons about the interaction of materials and building design can be learned from all the fires presented in this book. Some materials emit large quantities of fumes and smoke and should not be placed in buildings with building-wide systems such as ventilation, elevator shafts, plumbing, and telephone shafts. When a building has long corridors and/or long rooms, great care should be taken in the selection of finishings and furnishings. Very few architects and interior decorators consider the possibility of material or design interaction in a fire. Uninterrupted carpeting and wallcovering and large quantities of foam-filled, PVC-covered furnishings are common in corridors and are especially dangerous.

A special case of material/design interaction that should be considered is access to exits and firestairs. Placement of large quantitities of synthetics together can generate excessive quantities of lethal smoke that render exits and corridors to firestairs uninhabitable. Whether the material is placed directly in the corridor or exits or in spaces continuous with the corridors and exits does not seem to matter. The material that generated the lethal cloud in the MGM Grand Hotel Fire (see Chapter 6) was nowhere near the corridors. But the building-wide system, such as the elevator shafts, the ventilation system, and the seismic joints, made the space continuous and allowed the smoke to render the exits and corridors uninhabitable. The breaching of the firestairs made the whole building one big space for the smoke to fill. In the Ramada Inn Fire, the corridors became uninhabitable so quickly that the normal exits were unavailable even three to five minutes after the discovery of the fire. Too great a fuel load of synthetics can render the safety features of a building plan not worth the paper it's drawn upon. The access to exits becomes nearly impossible, and real lives are lost.

In a plastic world, realities blur: real safety features, real materials, real lives, and real deaths.

# CHAPTER 6

# MGM Grand Hotel Fire

Cities like Las Vegas, Atlantic City, and San Juan glitter and shine. Nothing is quite what it seems or resembles in these cities. A pseudo-Roman statue of Diana and a fawn that are in front of the Las Vegas Caesar's Palace aren't the glistening marble that they appear to be. Instead, they are plastic. Indeed, much of the city is composed of plastic, which makes it a disaster waiting to happen. And on November 21, 1980, it did.

When I went to investigate the MGM Grand Hotel fire, I wore old clothes, a hard hat, and a respirator that dangled around my neck when I wasn't inside the hotel. Male tourists thought I was a hooker with a clever specialty and offered some fairly high prices. You see, the city is given over to serving vices, and all unusual sights are quickly connected with these vices, even a woman in a hard hat and old boots.

Nothing is what it seems to be in the casinos, restaurants, and bedrooms. The "wood" furniture is rigid urethane or PVC. The "tapestry" wallcovering is plasticized PVC. The "glass" is polycarbonate or polymethylmethacrylate. In the walls and ceilings, much more plastic lies concealed in the form of electrical insulation, plumbing, and even parts of the air handling and HVAC (heating, ventilation and air conditioning) system. Metal trim, mirrors, velvet plush, and marble aren't what they seem. They are all plastic in this world of pseudo-pleasures.

The residents of Las Vegas look upon the tourists as boozers, gamblers, sexual adventurers, and gluttons. Tourist safety did not rank very high in the priority list of voters and members of civic associations until the MGM and Hilton Hotel fires. When income from tourism declined after these fires, and when the rest of the country asked hard questions about tourist safety in Las Vegas hotels, then remedial action came fast.

The economic status of the big casino hotels has often been pointed to as the rationale for the governmental agencies winking at the potential dangers in design, materials, and operations. Certainly, economics influences decisions. But the tourist life that forms the other part of the tradeoff between economics and life safety apparently weighs much less in the minds of our governmental regulators than it does in areas offering other kinds of recreational activities.

## THE HOTEL'S DESIGN

Like most of the major casino-hotels in Las Vegas, the MGM Grand Hotel occupied a large city block and rose over twenty stories above grade. We cannot assign a precise number of stories to these hotels because of the height of the casino and of the plenum above the casino. These two layers may each be two to four stories in height and cover more than the area of two football fields. No firewalls interrupt the huge spaces, so that any fire that would ignite in either would roar through the space, probably spreading to the other because of the many penetrations between them.

The plenum forms a work space for most casinos, including the MGM and connects with the casino. MGM, for example, used an "Eye in the Sky," an employee who sits above the casino and watches the games below. The plenum at the

MGM also held the air handling system controls and many fans, a grid of ABS and PVC drainage pipes (tons of plastic), and the vast electrical network, all wires insulated in plasticized PVC.

The casino below featured the gaming floor, with all the tables and rows of slot machines. Around the casino proper were the restaurants and bars.

Numerous paths linked the areas of the casino-hotel: the elevator shafts, the air handling system, the seismic joints, the electrical and plumbing systems, and the stairways. These paths could have been designed, built, and operated in a way that isolated the different areas in the event of smoke generation and fire spread. The seismic joints could have been capped at both ends, for example. The air handling system could have had automatic dampers, and shut off if it had a smoke detector within it. The fire stairs had been illegally penetrated in numerous places so that there were openings in the landings and other areas of the stairs that allowed smoke to enter the area. Thus, the design and operation of the hotel violated codes and practices for smoke control.

Let us return to the casino-restaurant first floor, where the fire started. The quantities of uninterrupted combustible materials in the habitable space and the plenum were estimated by Clark County investigators as, collectively, hundreds of tons. About ten tons of mastic (this particular glue was made from silicone plastic) held the processed wood ceiling tiles, which were the upper finish to the casino-restaurant complex. The "glass" of the slot machines, mirrors, and chandeliers was either polymethylmethacrylate (PMMA, an acrylic) or polycarbonate. The "wooden" tables were rigid polyurethane. Wallcovering, rigid molded furniture, and fake leather upholstery all contained PVC. The casino-restaurant area, which had no fire walls to break the space, provided uninterrupted access to an immense fuel pile in case of fire. Although much ado has been made about the lack of sprinklers in the hotel, it is possible that sprinklers would have been overwhelmed by the sheer concentration and pervasion of uninterrupted fuels. In its annual reports on fires causing over $1 million in damage, the NFPA has listed numerous fires in which sprinklers were overwhelmed by the sheer quantity of petrochemicals present under one roof.

Certain of the fuels, by themselves, would have generated a lethal cloud at casino level and, possibly, on the first floors into which the smoke traveled. These fuels are PVC, urethane, ABS, and the mastic. The full significance of the combination of all the fuels would require a full testing program to be explored. This testing program would have to be similar to that which the British Fire Research Center pursued to understand the dynamics of the Stardust Disco fire. The British combustion scientists set up a full quarter-room (a corner with its two walls and floor and ceiling), furnished and finished just as the disco had been. Then, they ignited a fire and let it burn.

PVC, which decomposes readily, existed in the same environment in the casino as ABS, which burns readily and emits hydrogen cyanide, and as PMMA, which burns readily and emits methylmethacrylate, which is its monomer and an irritant and nerve poison. In general, combined dosing has proven worse than single-type dosing, toxicologically. No lab tests have ever been performed to expose animals to the mixes of acids and organics that would have been present in smoke such as that generated by the MGM fire. As mentioned in Chapter 5, these products are tested individually, not in combination.

## THE FIRE

All of the parties involved in the litigation agreed that the fire originated in the electrical system in the back wall of the kitchen/busboy station of the deli restaurant. The burn damage was very strange, because most of the objects in that room had only smoke damage. Paper announcements hanging on the walls showed no char at all. Loaves of bread sitting on a cart weren't even toasted. Plastic dishpans outside the limited area of direct fire damage weren't even melted. From the pattern of flame damage, the path of fire spread could be pieced together.

Some flames did shoot out directly from the wall. The major path of fire, however, led up the wall to the plenum by way of the electrical installation. The plastic pipes in the plenum, as well as the electrical insulation, then carried the flames across the plenum. The fire poured down into the deli restaurant from the ceiling about twenty feet from the wall of origin.

Before the fire began its rapid and unstoppable spread, a guard noticed smoke coming from the wall. Very shortly after he sighted the smoke, he heard a "pop"—a small explosion—and saw the flame shoot out from the wall. He ran to sound the alarm, but the fire began rolling through the plenum before the firefighting units could respond.

The pipes in the plenum had given the fire a directed energy and transformed it into the much-described "fireball." This was a front of fast moving flame that crossed the casino (a distance of about 200 yards) in about five minutes. It was a coherent mass of flame moving in one direction, and actually burst out of the Flamingo Avenue doors, setting the canopy on fire and damaging the first-to-arrive fire engine that had parked there. It did not consume the entire casino-restaurant area, but a distinct pathway within it. Lounge seats within its path were completely destroyed except for their bare springs, but lounge seats only a few feet from those destroyed were completely undamaged by flame.

Within the pathway of the fireball, most of the damage to the carpeting was surficial, except for a peculiar gridlike pattern of deeper burns. This pattern reflected the layout of the ABS pipes above in the plenum and showed the impact of the added heat from above. Thus, the ABS was responsible not only for its own smoke but also for the added smoke and fumes from the deeper burned objects below in the casino. These kinds of interactions have never been considered when standards and codes have been made, but may contribute significantly to smoke toxicity and toxic hazard.

The fire occurred at 7:30 in the morning, a time when few people in Las Vegas hotels are awake. In fact, few people were on the casino level at the time. Even fewer were in the direct path of the fire. Much has been made of the speed and heat of the fireball. It is estimated to have reached 1,200°F. For a building fire, this is a rather standard temperature. Firefighters are told in their training that building fires reach 1,000°F within five minutes of ignition. Other fires have been much hotter; the New York Telephone fire reached 3,000°F and buckled structural steel beams. This fire was also not unusually fast as far as hotel fires go. The Stouffer's Inn fire (described in Chapter 7) was at least as fast, if not faster. The flames were not that unusual in speed or intensity, and did not directly kill many people.

What was unusual about this fire was the smoke: its quality, quantity, density, and the number of people it killed.

The most striking fact about the MGM fire, a fact repeated time and again by newscasters, firefighters, and fire scientists, concerns the distribution of the fatality victims: the great majority of them (sixty-one out of eighty-five) died on the nineteenth through the twenty-sixth floors of the hotel (the twenty-sixth floor was the top floor). These victims were as far from the fire as they could be and still be in the building. The smoke had risen to the top floor, accumulated, and spread downward and up out of the building top.

The avenues of this spread to the top floor included the air handling system, the elevator shafts, the seismic joints, the fire stairs, the electrical and plumbing systems, and even the broken windows on the windward side of the building.

The air handling system did not have a smoke detector connected to automatic dampers. Smoke had a clear path through the ducts. In addition, the PVC (!) pneumatic tube that connected the control panel to the fans melted in the early stages of the fire, being located in the casino plenum. PVC should never be used for this purpose. Control over the fans was lost when the PVC tubing melted, and the fans continued to push smoke around the building. In the rooms in the tower, many of the occupants had tried to diminish the amount of smoke entering their rooms by placing wet towels over the air outlet from the supply duct. One of the first cues that the air handling system was an avenue for the smoke came when investigators noticed the towels over the air supply grating.

The elevator shafts and the seismic joints could have been constructed so that they would not act as chimneys. The seismic joints, especially, could simply have had a cover over their bases instead of being open to the casino plenum.

The electrical and plumbing penetrations were marked by black soot. These are supposed to be completely firestopped so that they will not provide an avenue for firespread. Firestopping would also have reduced or eliminated the spread of smoke. The fire stairs were illegally breached with plumbing penetrations and proved a deathtrap for several fatality victims who took refuge in what should have been a safe, smoke-free area.

Thus, the smoke and heat rose to the top of the building. The smoke was so voluminous that even the towering pillars that were given off by the fans at the building top could not relieve the pressure sufficiently to keep the smoke from spreading downward when it filled the top floor and had nowhere else to go. Even the windows that were broken to obtain fresh air proved to be avenues of smoke when the wind shifted and blew a column of smoke back into the building.

## THE VICTIMS

This cloud of smoke caused eighty-five fatalities. The great majority were found in the top floors (floors nineteen through twenty-six) and near the avenues of smoke movement. Table 6.1 shows the location of the fatalities.

The location of death may not be the place where the lethal dose was received. People may have received an initial lethal dose and then moved around for a while to try to save themselves. However, the locations of the bodies generally neighbored the major avenues of smoke described in the

**Table 6.1.** Location of the Fatalities.

| Floor | Men | Age Range | Women | Age Range |
|---|---|---|---|---|
| 26 | 0 | | 1 | 31 |
| 25 | 1 | 48 | 5 | 19–60 |
| 24 | 6 | 22–46 | 4 | 23–55 |
| 23 | 8 | 37–59 | 5 | 36–59 |
| Stairwell | | | 1 | 53 |
| 22 | 2 | 44, 69 | 2 | 53, 71 |
| 21 | 6 | 25–59 | 2 | 33, 35 |
| 20 | 8 | 22–72 | 6 | 30–70 |
| 19 | 2 | 36, 69 | 2 | 34, 63 |
| Total (M) | 33 | Total (F) | 28 | |

previous paragraphs. One of the eighty-five deaths can be indirectly attributed to the smoke. A woman jumped from a high floor in a panic trying to escape the smoke.

There were over 500 injuries. The data on where these people were at the time of injury are not available. They certainly occupied a broader range of floors than the fatality victims. They also represent a broader range of exposures. Some got out of the hotel within only five minutes of the arrival of the fire department, and some were rescued after two or three hours on the roof or a mid-level floor.

The fatality victims ranged in age from twenty to seventy-two. An almost equal number of men and women died. Most of the victims were between thirty-five and fifty-five years old.

Obviously, the victims on the upper floors had no direct contact with the flames and died of smoke inhalation only, with the exception of the woman who jumped from her window. Although several of the victims who died on the first floor had burns that were sometimes extensive and severe, the general consensus of the pathologists who examined the bodies was that most of these people died of smoke inhalation and were burned after death. Even those who may have died partly from burns appeared to have first been incapacitated, and showed signs of sudden fall-down.

These poor corpses were poked at, analyzed, chemically separated, and written up by many different health professionals from several different agencies. Their tissues were distributed on both the East and West Coasts. Blood, bronchi, and other samples were sent all over the country.

The blood was analyzed for carbon monoxide and cyanide as well as alcohol and drugs. As would be expected in Las Vegas, some of the samples showed small residues of alcohol, but the great majority did not. Any alcohol that was consumed the previous evening was metabolized by the time of the 7:30 a.m. fire. The majority of the victims showed lethal levels of carbon monoxide in their blood, but a significant proportion (47 percent, according to the NBS report), showed sublethal levels. This tells us that other chemicals in the smoke contributed to the fatalities.

Cyanide was present in the blood of a number of victims. Unfortunately, the methods for preserving the bodies and the blood samples and for analyzing them for cyanide were not good enough to allow reliable quantification of the cyanide. We cannot say which of the victims had lethal levels of cyanide in their blood. All we can say is that some of the victims inhaled smoke containing cyanide. This information supports the notion mentioned above that other chemicals besides carbon monoxide contributed to the fatalities.

We use the National Bureau of Standards carbon monoxide analyses, rather than the analyses of the Clark County coroner or Johns Hopkins Medical School. The office of the Clark County Coroner had taken certain shortcuts and produced unreliable results. The Johns Hopkins results were not reported *in toto,* and the piece of paper shoved under the noses of expert witnesses during deposition showed only the average of several different analyses of the same blood samples. We had no idea of the reliability of the analyses because the raw numbers were never produced for review. Therefore, the only analysis to be relied upon is that of the National Bureau of Standards. The Johns Hopkins results were being supported during litigation by the defendants who manufactured and distributed the various plastic products that were consumed in the fire and that contributed to the smoke and fumes.

Besides the blood chemistry, the pathologists who performed the autopsies made a number of important observations about the organs of the victims. One of the frustrations they had in trying to obtain carbon monoxide concentrations in the blood was that the red blood cells in a few of the victims simply disintegrated. The destruction of red blood cells has been seen in victims of other plastics fires and in lab animals exposed to PVC fumes. Hydrogen chloride destroys oxygen-carrying hemoglobin, the protein that forms the major content of red blood cells.

The autopsy protocols typically listed a number of observations about common conditions in the victims: irritated eyes, reddened lining of the respiratory tract, large quantities of soot throughout the respiratory tract, frothy bloody liquid in the lungs, lung tissue destroyed, influx of a large volume of blood into the blood vessels of major organs (liver, kidney, brain, lungs), and copious mucus from nose. A few victims showed bleeding from ears and nose, originating from the pharynx. Some victims showed signs of cyanide effects, with ashen gray coloring to the skin and lining of the mouth. Carbon monoxide alone would not have caused these effects, although signs of carbon monoxide were seen, such as the typically cherry-red coloring of the lips and some areas of skin. Cyanide was present in at least some areas of the building, and corrosive irritants were present in all the areas where victims were found.

Two scientists performed special work on tissues from the victims. Dr. Oscar Hunter, a nationally known pathologist, looked at respiratory tract tissue and noted signs of irritation where soot had been deposited. He saw definite blistering. Dr. Merritt Birky, then head combustion toxicologist with the National Bureau of Standards, chemically analyzed the soot found in the bronchi of two victims, the soot found in the upper floors, and the unburned samples of products present in the casino, products that had burned during the fire.

The elements found in both samples of soot correlated fairly well. In trying to trace the soot back to the products in the casino, the NBS team had some difficulty, because some of the products contained the same elements. For example, the vinyl wall covering contained kaolin as a filler, a clay that has calcium in it, but the mastic of the ceiling tiles also contained calcium. Some of the elements appear to have come from PVC products: antimony, zinc, and lead, for example. Many forms of plastic may contain a bromine-based fire retardant, so finding bromine in the soot did not help to trace the soot back to any particular product, but does trace the soot to synthetic polymers as opposed to wood. What we can conclude is that the synthetic polymeric products in the casino were the source of the soot found in the rooms and in the victims' bronchi, because wood does not contain these elements in large quantities. We can also conclude that at least some of the soot came from the PVC products, specifically. Sixty-one people died twenty stories above the fire from soot and fumes given off by the burning plastics in the ground-floor casino.

## THE CONSEQUENCES

Because of the large number of plaintiffs in the MGM fire litigation, the court asked that a limited number be chosen as example plaintiffs and that their deaths or injuries be looked at in detail. This decision led to a limited number of the injured entering the data base, although the autopsy protocols of all fatality victims did comprise part of the data base.

My data base in injuries consists of medical records and/or health status questionnaires from fifty-four survivors. Everyone in the data base suffered from smoke inhalation effects and psychological stress from the disastrous nature of the event. No one had involvement with the flames or burns from the fire itself.

Nearly all the people in the data base suffered from some form of respiratory injury, and initially these injuries showed great variety. (Refer to Table 6.2.) The injuries included sensitization and asthma, obstructive lung disease, loss of immunity from the respiratory tract, sinus troubles, dry throat and frequent sore throats, and frequent phlegm production. Other injuries listed or documented among a high proportion of the injured included dry, sensitive skin; shifts in sleep pattern; menstrual problems (very heavy flow, pain, and irregularity); excessive sweating; poor circulation in hands and feet; muscle spasms in the jaw or lower back; and depression/anxiety. Table 6.3 lists non-respiratory injuries by physiological system and the number and percent of the people who had them. The pattern of injury differs greatly from that caused by carbon monoxide alone or by the combustion of wood or paper.

Besides the group pattern shown in Table 6.2, certain individuals experienced special disabilities from the smoke. The muscle spasms mentioned above show that certain parts of the nervous system that cause involuntary movement of the normally voluntary skeletal muscles were affected. Both Parkinsonism and epilepsy are based in these parts of the nervous system. In addition, sufferers of Parkinsonism and

**Table 6.2.** Respiratory Complaints of 58 Survivors.

| Complaint | Number | Percent |
|---|---|---|
| Frequent sore throat | 30 | 51.7% |
| Hoarseness | 26 | 44.8% |
| Sinusitis | 23 | 39.7% |
| Sensitive to dust/smoke | 44 | 75.9% |
| Coughs | 27 | 46.6% |
| Wheezing | 26 | 44.8% |
| Phlegm production | 29 | 50.0% |
| Shortness of breath | 38 | 65.5% |
| Bronchitis | 21 | 36.2% |
| Frequent upper respiratory infection | 25 | 43.1% |

**Table 6.3.** Non-Respiratory Symptoms of 57 Survivors.

| Symptoms | Number | Percent |
|---|---|---|
| *Circulatory system:* | | |
| Developed heart problem | 12 | 21.1% |
| Hands/feet fall asleep | 39 | 68.4% |
| Hands/feet easily cold | 22 | 38.6% |
| Mottling or blue hands/feet | 8 | 14.0% |
| Blood pressure change | 20 | 35.1% |
| *Neurological/psychoneurological:* | | |
| Headaches | 34 | 59.6% |
| Change in sleep pattern | 42 | 73.7% |
| Memory lapses | 32 | 56.1% |
| Irritable | 43 | 75.4% |
| Difficulty learning | 14 | 24.6% |
| Change in perception abilities | 31 | 54.4% |
| Coordination decline | 15 | 26.3% |
| Dizziness | 27 | 47.4% |
| Disorientation | 20 | 35.1% |
| *Kidney/bladder:* | | |
| Frequent urination | 15 | 26.3% |
| Pain/burning during urination | 12 | 21.1% |
| Lower back pains | 27 | 47.4% |
| *Skin changes:* | | |
| Acne-like breakouts | 13 | 22.8% |
| Rashes | 12 | 21.1% |
| Burns from the smoke | 4 | 7.0% |
| Infections | 6 | 10.5% |
| Skin growths (warts, moles) | 3 | 5.3% |
| Skin dry, sensitive | 14 | 24.6% |
| *Reproductive:* | | |
| New menstrual difficulties | 11 | 84.6%* |
| Miscarriage | 2 | 15.4%* |
| Hysterectomy | 2 | 15.4%* |
| Dilation and curettage | 1 | 7.7%* |
| *Psychological:* | | |
| Nightmares | 42 | 73.7% |
| Depression | 40 | 70.2% |
| Guilt | 15 | 26.3% |
| Anger | 17 | 29.8% |
| Change in relationships | 24 | 42.1% |

**Table 6.3.**—*Continued*

| Symptoms | Number | Percent |
|---|---|---|
| *General well-being:* | | |
| Change in appetite | 16 | 28.1% |
| Less endurance | 31 | 54.4% |
| General weakness | 24 | 42.1% |
| Ceased/decreased activity | 17 | 29.8% |
| Lack of sexual desire | 14 | 24.6% |
| *Other complaints:* | | |
| Sensitized eyes | 11 | 19.3% |
| Dizziness with position change | 8 | 14.0% |
| Heavy sweating | 7 | 12.3% |
| Drink more water | 3 | 5.3% |
| Tremors, spasms, clenching | 6 | 10.5% |

*Percentage is that of the women of reproductive age in the data base, not of all 57 people.

The following affected less than 5 percent: liver dysfunction, nausea, neck pains, impotence, pigmentation of skin, loss of hair, ears hurt, swollen glands, onset of epilepsy, exacerbation of Parkinsonism, return of menses to elderly woman, swollen testes, prostate infections.

epilepsy generally have low levels of one or more of the nerve chemicals called **catecholamines**. Depressed persons also have low levels of certain catecholamines. One injured person had Parkinsonism even before the fire, but required a much greater amount of medication to control his condition after the fire. A second person became epileptic after the fire and suffered from grand mal seizures. These two individuals are index cases of the neurotoxic nature of the smoke. The muscle spasms and possibly the depression are confirmation of this.

Besides the physical injuries and psychological distress, social interactions appear to have been affected by the experience of the fire. Marriages became troubled and were dissolved. The divorce rate among the survivors was very high. This high rate of divorce is also seen in other disasters besides fires, and reflects personality changes and relational changes that often result from having risked death and knowing that others died in the same event. Besides divorce, another socioeconomic consequence of the fire was the decline of several businesses that had sent key people to conventions or seminars that were held at the hotel. These people stayed overnight at the hotel, and several within the same office may have suffered personality changes. Individual productivity typically declined, but group interactions also suffered so that decisions could not be made, consensus could not be reached, and actions could not be taken. At least two middle-sized companies suffered greatly and became much smaller when several key managers could not perform.

The firefighters who fought the MGM Grand Hotel fire also suffered injuries. National Institute of Occupational Safety and Health (NIOSH) conducted a particularly incompetent hazard evaluation of the effect of the fire on these firefighters. This agency actually obtained positive results from their case/control study. In this study, these firefighters were matched with a control group and compared with respect to certain health complaints. These firefighters had certain respiratory problems, sleep difficulty, irritability, and depression in greater proportions than the control group. The NIOSH conclusions were incredible: the firefighters had these problems because they are firefighters, not because of the fire. These problems were identical to those seen in the injured civilians. NIOSH carefully avoided looking at data on the civilians to compare them with those of the firefighters. During the Reagan administration, NIOSH produced an astonishing number of studies showing either no effect of job-related conditions on worker health or effects caused only by interaction between workplace exposure and the worker's personal habits. In other words, if those darned workers would only stop smoking and drinking, all these health problems would go away. It's all their fault. Like so many other worker groups, the firefighters got shafted by an agency that is supposed to help them.

The firefighters' local union raised issues of manning and training in connection with this fire. With ordinary building design and content, these issues would be important. With the design of Las Vegas casinos, and their contents being what they are, an expansion of the work force would still not be enough to overcome these factors, nor would any level of training. These quality-of-service issues do pertain to office buildings, apartment buildings, and other buildings of plastic structural elements and furnishings interrupted by firewalls. These issues must be considered for "normal" buildings, especially in "normal" cities. The Las Vegas fire-fighter has two cities to serve, the residential one and the resort area. There is hope that an adequate number of units and manning of units can control the fires in the first; there is no such hope for the second. Both firefighters and civilians will continue to face injury and death because of the design, construction, and content of the casino hotels.

## THE INVESTIGATION

Much of the evidence summarized above implies conclusions about the various factors that contributed to the deaths and injuries in the fire: the physical nature of the soot (its chemistry and quantities), the locations of the fatality victims, the autopsy results, the medical records of the injured survivors and their health status questionnaires, and the events of the fire as seen by witnesses. In science, there is a basic rule called *Occam's Razor,* which states essentially that explanations for observations must be as simple as possible while still covering all the observations, and this is what we must do here.

One of the most unambiguous lines of evidence is the quality and quantity of the soot produced by the fire. Most of the same elements found in the soot sampled from the rooms of the upper floors were also found in the soot in the bronchi of the fatality victims. This tells us that the victims breathed in the products of decomposition from the same materials that gave rise to the soot in the rooms. It does not tell us that this soot actually killed them, but that it was present in their bodies at the time of their deaths. We can trace this soot back to many of the products that were burning in the casino area, the mastic and the PVC in particular. Other products congruous with the elements found in the soot include the ABS plumbing pipe and the urethane foam in the cushioned seating used in the casino. Qualitatively, we can say that the soot found in the upper floors and in the victims' respiratory tracts reflects the products that burned in the casino area. The indications of inflammation and blistering found in the tissue apposed to the soot deposits in the bronchi also tell us that this soot contained corrosive irritants capable of producing inflammatory reactions.

Let us look at the soot elements in more detail. They are listed in Table 6.4 along with the possible products that contained them and that we know burned. Not all of the elements were found in both victims. As you can see, several of the elements could have come from more than one product. When you put them all together, you can see that PVC must be the source of a number of them. Also, the quantity of mastic that burned is consistent with contribution to the high quantity of silica, aluminum, magnesium, and calcium, only part of which could have been supplied by the plasticized PVC wallcovering. When we add the knowledge that the smoke also contained cyanide, then several of these elements, as indicated in the Table, most likely came from the

**Table 6.4.** Elements Found in Soot of Bronchi from Two Victims.

| Elements | Possible Source |
|---|---|
| Potassium | Common in soot of nearly any fire |
| Iron | PVC, mastic, ABS |
| Chlorine | PVC plus any product with chlorinated fire retardant |
| Zinc | PVC |
| Lead | PVC |
| Bromine | Nearly any product with brominated fire retardant; could have been any polymer |
| Phosphorus | Urethane foam |
| Sulfur | Rubber |
| Nickel | Mastic, PVC |
| Calcium | Mastic, filled PVC |
| Copper | Electric wiring, copper decorations |

ABS pipe and urethane furnishings. Unlike the soot samples in the Stouffer's Inn case, these samples can only confirm other evidence, because too many products burned in the MGM Grand Hotel Fire and contained the same elements for any conclusion to be based solely on soot elements.

Other physical evidence that may teach us something includes the burn pattern, the smoke path, and the air handling system. Of all the fires in my file, the MGM had the most striking burn pattern. The fire front had such direction and impetus that it divided the casino into three sharp zones, one of charring and destruction and two on either side with little or no fire damage. The fire dynamic had already gained this momentum by the time it burst out of the ceiling and into the restaurant part of the deli. The plastics hidden in the wall and ceiling determined thc fireball speed and direction in only a few minutes. The fire front raced so rapidly through the casino that, even within the charred area, the carpeting showed mere surface burns, although the furnishings in the hotter upper layers of the atmosphere were completely destroyed. As mentioned before, the carpeting showed a peculiar gridlike pattern of deeper burns within the surface charring, a reflection of the furiously burning ABS drain pipe in the plenum above. The hidden plastics continued to pump energy into the fire below. The heat of the fireball passed through the casino so rapidly that it was confined to the freely moving layer of atmosphere above the floor. The floor protected the carpeting somewhat by a boundary layer effect. (The **boundary layer** is the atmosphere immediately adjacent to the solid.) But the heat release of the ABS piping above affected even the boundary layer. The lesson from the burn pattern is that energy interactions will occur when large quantities of many kinds of plastics decompose and burn in one building unit. These energy interactions will determine the dynamics of the fire spread and the quantity and quality of smoke and fumes generated. The energy interaction determines both the dynamics of the fire and its spread and the dynamics of the toxic exposures from the smoke generated from the fire. Because of gaseous flows, initial smoke exposures generally precede any exposure of significance to flame.

If this fire had occurred later in the day or in the evening, hundreds of people would have perished. Fortunately, ignition occurred early in the morning when few people were on the casino level.

One major factor leading to the deaths and injuries of the guests in the rooms above was the sheer quantity of smoke which sought every possible pathway from the casino level. The design and construction of the hotel afforded numerous pathways: the smoke traveled up through the siesmic joints, elevator shafts, stairways, air circulation ducts, and utility penetrations. Even the windows on the windward side of the building became smoke pathways. The building design and construction, including the makeshift illegal changes such as penetration of the fire stairs by pipes, gave smoke the opportunity to reach the upper stories and to travel downward through the upper parts of the hotel. But the sheer quantity of the smoke and its toxicity were such that someone would have been injured by exposure even under the best design and construction practices. The design and construction of the MGM Grand Hotel, however, increased the number of people that were killed and injured.

The filters on the air handling system showed what kinds and quantities of smoke raced through the hotel. The filters were overwhelmed—they were completely black and clogged. There were at least two layers of filters: one by the fans on the way to the rooms and one at the room level (each room had a filter). Both layers of filters were black. In the rooms where there were fatalities, the hotel tried to tamper with the evidence by replacing the room filters. We found loosened screws on the corners of the grill to the air supply vent and pieces of tissue or paper toweling used to clean the grill. When we looked at other rooms on the same floors, we found blackened filters and grills. The air had circulated during the entire fire, but the filters could not save anyone. Indeed, the air had circulated because the PVC pneumatic control tube had melted during the early stages of the fire in the casino plenum, leaving the fans without control at all.

Even twenty stories above the fire, with all the dilution of the smoke, people could attain fatal levels of carbon monoxide. This is a remarkable fact. Everyone has been busy pointing out the fact that a certain percentage of victims had sublethal levels of carbon monoxide in their blood, in order to implicate other chemicals, such as the acids and cyanide, as proof of the toxicity of the plastics. A very important and overlooked fact is that plastics generally produce larger quantities of carbon monoxide per unit volume than natural polymers produce. And, they produce much larger quantities of carbon monoxide than inorganic solids, such as metal and glass, for which plastics often substitute. (Other chemicals besides CO can and do cause deaths in plastics fires, but just on the basis of CO production, plastics are more dangerous than wood or cotton, and much more dangerous than metal or glass.) A second very important fact is that plastics are generally present in much greater weights than natural polymers. Silk, wool, cotton, or even wood are usually not present in quantities of several tons and would not produce the huge quantities of carbon monoxide that plastics, which are often found in massive quantities, produce. The high concentrations of carbon monoxide produced by plastics means that the potentially lethal area of smoke and fumes is much greater even if only the carbon monoxide, and not the other gases and soot chemicals, is considered.

Of course, the other data from the autopsy protocols and medical records show the presence of acid gases, cyanide, and hydrocarbons in the fumes and smoke. The lung damage seen in both the fatality victims and the injured survivors was typical of inhalation of acids. The reactive airways seen in the survivors also indicated inhalation of acids.

Several of the common health problems of the survivors—the skin sensitivity and dryness, the neurological signs, and the problems with microcirculation in extremities—could be assigned to either the acidic irritants or to the hydrocarbons. Other problems have typically been caused by hydrocarbons, especially chlorinated hydrocarbons: uterine dysfunctions, excessive sweating, muscle spasms and shaking, and skin rashes, acne, and discolorations.

Some of the strongest symptom patterns were psychological. Depression, irritability, nightmares, inability to concentrate, and relational problems with friends and family were common in survivors. Irritants and hydrocarbons both have been found to influence psychological function, especially through the catecholamine systems. Disasters, however—especially disasters in which people have died—have also been found to cause post-traumatic stress syndrome. Many who have been through these large fires complain about flashbacks triggered by smelling very irritating smoke. The chemistry of the smoke and the life/death nature of the disaster may be so completely intertwined that parceling out the effects of one or the other on the psychological functioning of the survivors may be an impossible and unproductive task.

## THE LESSONS

Let us summarize the lessons of the MGM Grand Hotel Fire, lest 85 people should have died without leaving a legacy.

### Lesson 1

The quality and quantity of fuel load, especially fuel load uninterrupted by fire walls or other firestops, must be considered when a place of public assembly or public accommodation is on the drawing board. Estimates of the quantity and toxicity of the smoke and fumes that would possibly be generated by this fuel load should be part of the basis for choice of materials and for material usage. If we were to propose building a nuclear reactor, we would have to develop a worst case scenario. Now this kind of thinking ahead is even being applied to chemical plants and warehouses. Certainly, the plastics produced by these chemical plants and stored in these warehouses, plastics that are related both by lineage so to speak, and by structure to pesticides, require emergency response planning and designs that minimize exposure.

**Lesson 2**

Building design can interact with fuel load in disastrous ways. The bigger the building, the bigger the potential disaster. The larger and more toxic the fuel load the bigger the disaster. Also, carelessness in building design and construction will augment any problems in a fire from a nontraditional fuel load. The less traditional the fuel load, the more necessary that every feature of the building be designed to follow specifications that are accumulated through knowledge about safety and that all methods of keeping smoke and fire from spreading and of giving occupants early warning be used. This may even mean extra-dense sprinkler systems; sprinklers and smoke alarms in the hidden spaces, such as ceilings and walls; and methods of pressurizing both corridors and stairs so that escape routes are free of smoke. It also includes such accepted methods as automatically operating dampers in the air ducts. It may be that such heavy use of plastics as seen in the MGM Grand Hotel requires exorbitant investment in order to partially overcome the fundamental properties of the materials. The cheapness of these materials is a myth, an illusion broadcast by the industry. These materials cost us all plenty.

**Lesson 3**

This lesson destroys the illusion that if we follow a host of little "safety tips" in case of a hotel fire, we will come out alive. Many of the guests followed these safety tips. Some had even studied NFPA pamphlets about hotel fire safety. Three guests who were interviewed had read the NFPA pamphlet after previous hotel fires had made them concerned. But there was no way to be safe in the face of the huge and toxic fuel load, the pathways of smoke into the rooms, and the contamination by smoke of the supposedly "safe" stairs. People who stayed in their rooms died. People who left their rooms died. People who put wet towels over their air supply vents died. To mislead the public into thinking that they could survive a disaster like the MGM Grand Hotel fire would be like giving safety tips on how to survive in Pompeii when Mount Vesuvius blew. The little pamphlets place a great burden on citizens by implying that if you do all the right things and are a responsible, moral person, you will survive a hotel fire. If you don't survive, according to the ethos of these pamphlets, you didn't do the right things or you were drunk or otherwise immoral. Of course, we all have the responsibility of knowing how to behave in a way that maximizes our safety in a disaster of a common nature, but there are other parties in this responsibility to keep us alive and well; government, quasigovernment, and industry. Their failures kill us.

*CHAPTER 7*

# The Stouffer's Inn Fire

The town of Harrison, New York is in wealthy Westchester County, just north of the Bronx. During the 1970s, Harrison grew rapidly. New homes, office buildings, restaurants, and other businesses sprung up, but Harrison did not update its building and fire code. The town retained a code from the 1950s. Under this antiquated code, the main building of the Westchester Stouffer's Inn was issued a building permit in 1975, and a certificate of occupancy in 1978.

The main building contained meeting rooms, a ballroom, and other facilities for groups, but no guest rooms. It was sometimes called The Conference Center and classified as a place of assembly. Three stories tall, it had the code classification of fire resistive, i.e., built of noncombustible materials.

Figure 7.1, taken from the NFPA report on the fire, shows the arrangement of the third floor, where on December 4, 1980, a deadly fire broke out. Figure 7.1 also indicates that the fire originated where three corridors met, raced down the corridors, spread smoke widely, and killed twenty-six people. This fire ignited at about 10:15 a.m. and was discovered at about 10:20 a.m. By the end of 1983, the Stouffer's Corpo-

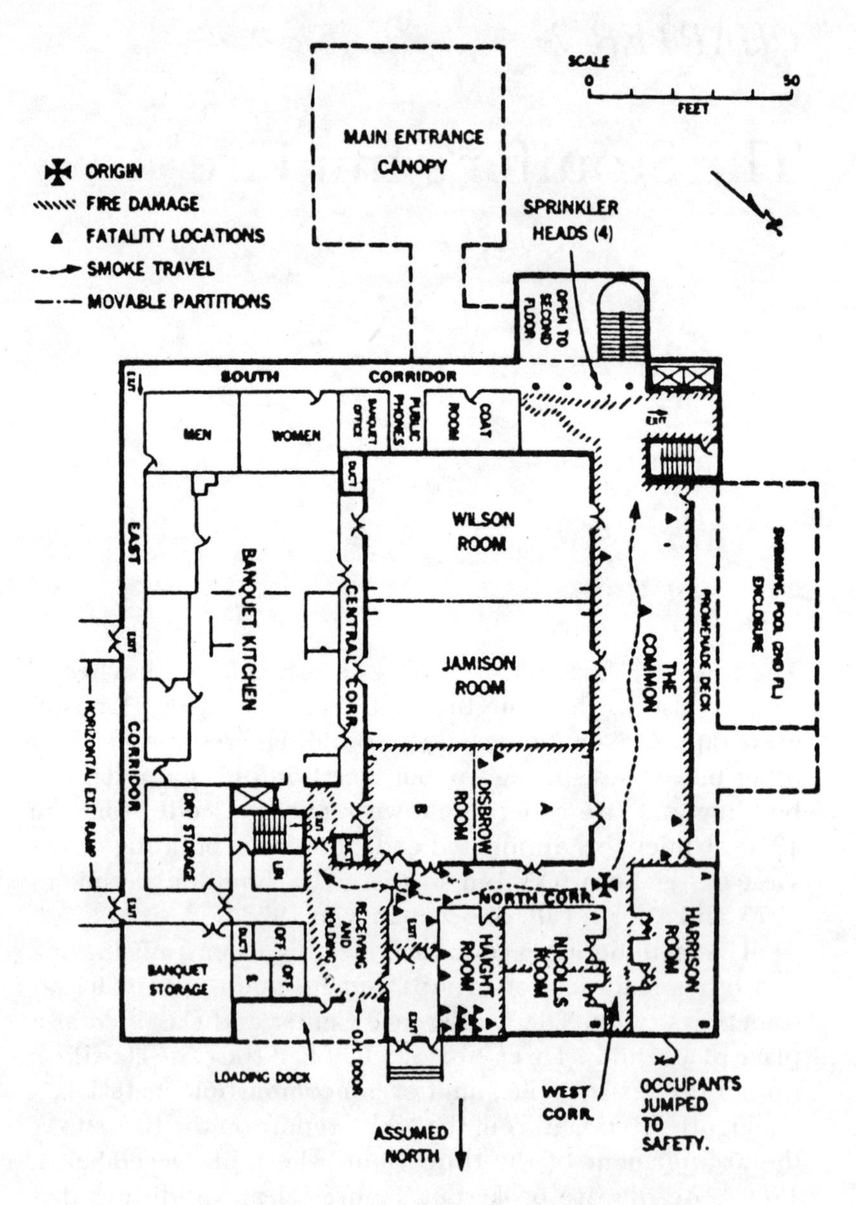


Taken from NFPA report.

**Figure 7.1.** The Layout of the Third Floor of Stouffer's Inn and the Geography of the Fire.

ration and a series of subrogation defendants agreed to a settlement of approximately $48 million with the estates of the fatality victims and with the injured guests.

## THE BUILDING'S DESIGN

As indicated by its classification "fire resistive," the basic construction material of the building was noncombustible: steel, concrete, and glass. Smaller quantities of aluminum and stone were also used. Because of this classification, the building had few sprinklers. However, as is common to motels and hotels of the modern "fireproof" construction, the rooms and corridors held a dense fuel load in the form of synthetic furnishings, finishings, and decorations. The town, like many others, had no regulations governing these products, which aren't considered part of the building systems.

The corridors and the "Commons" had uninterrupted carpeting of a nylon/wool blend. The walls of the corridors and the meeting rooms were covered with plasticized PVC wallcovering of the standard composition (approximately 40 percent plasticizer, 60 percent PVC). The carpeting and the wallcovering formed the two major fuels during the early stages of the fire, when guests and employees were still in the building. These finishings also provided uninterrupted, relatively thin avenues of fuel along which the fire could easily travel.

The other furnishings and decorations certainly added to the heat and smoke, but would not have led to the tragic outcome if they alone had been set on fire. The rapid spread and density of the smoke and the rapid spread of the fire depended on corridors that were lined with combustible, toxic finishings. Furnishings that contributed their toxic products included a foam-filled Naugahyde-covered couch, a PVC artificial Christmas tree, a coat rack with coats on it, several small tables, polyester draperies, and some PVC-covered chairs. A wooden piano also burned, but not during the crucial early stages.

There were a total of 95 people in six of the meeting rooms on the third floor. These rooms were the Wilson Room (42), Disbrow A (22), Disbrow B (6), the Harrison Room (11), and Haight Room (13), with the Nicolls Room A being used for electronic equipment (refer to Figure 7.1). In addition, Stouffer's employees were in the kitchen and other places. As can be seen from Figure 7.1, the occupants of these rooms had very limited options for exits. The people in the conference rooms were executives and managers from various brewers attending a United Brewers Academy seminar and Nestles, Pepsico, General Foods, and Arrow Electronics. Besides the injuries and loss of life, the fire resulted in the demise of Arrow Electronics.

## THE FIRE

The origin and spread of the fire, although graphically shown in Figure 7.1, comes to life from the eyewitness testimony in the criminal court case. The quotes below are from the digest of the criminal trial produced by the plaintiffs' counsel. They are from four different witnesses.

> She saw a fire right in the center of the carpet where her hallway intersected with the other hallway. . . . It appeared to be a bonfire sitting in the middle of the carpet with space to the left and right. It appeared to be about five feet high and about five feet wide. It seemed to her that just the carpet was burning. . . . It seemed like a long time while she stood and watched. . . . Then it occurred to her it was out of control. It started to spread as she stood there. It spread both left and right, but faster to her left which was toward the Haight hallway. . . . She decided not to go back into the room to get her materials. At that point she started to run and she ran to the right side of the fire as she passed it. . . . The smoke was very heavy at that point; it had gotten much darker and difficult to breathe. She stopped breathing and kept running until she passed the smoke. As she reached the top of the stairs, she looked back. The smoke was clear

> at that point and she saw that the flames covered from the windows all the way past where she could not see them into the Haight hallway and that they were reaching up to the ceiling. The entire width of the Commons was in flames up to the ceiling.
>
> Smoke was coming into the room like a screen. It was black. When he got to the Haight hallway and looked to his left, he saw flames and smoke and it appeared to him that the flames were coming from the ceiling. It appeared to be like a fireball coming along the ceiling and being blown at them. He told people that he thought it was the wallpaper that was burning and that he couldn't imagine the wallpaper burning that much.
>
> The first thing he saw was a small bundle of fire which was about three to five feet in width and about three feet in height, as if someone was stoking a fireplace.... He then left the landing after speaking to a bellman and ran up to the top of the staircase again. This time it had grown much larger, the height being six to eight feet and much larger.... He made eye contact with Mr.—— after having gone down to the landing again. He then came up the stairs again and observed the fire for the third time. He was staying there trying to decide what to do when smoke started coming down the hallway toward him at a tremendous pace.... When the smoke came gushing down the Commons, and from the top of the ceiling, it was billowing down, gray smoke from the top.
>
> Less than two minutes went by from the time Raj yelled fire to the time he took the photograph with the flames coming out of the window (and Raj yelled fire when it was a very small fire).

The fire grew slowly and radially for its first 3–5 minutes and then spread with high acceleration down the corridors, especially the corridor to the Haight Room.

From the diagram of the third floor plan taken from NFPA (1981) (Figure 7.1) we can produce data on the location of the fatalities. See Table 7.1.

Clearly a disproportionate number of the fatalities occurred in a small part of the total area affected by fire and/or smoke: 19 died in or near the Haight Room (73 percent), but this area constituted only a small percentage of the total area affected by fire and/or smoke. Ninety-five occupants had been present, so slightly over one-quarter (27 percent) of those present died. Another 24 (25 percent) were injured, for a total of over half of the occupants physically affected by the fire.

The eight fatalities in the North Corridor had all been occupants of the Disbrow A room and had time only to enter the corridor. In contrast, all but one of the occupants of the Wilson Room were able to exit the floor without fatality. The one who died was the last person to leave the room and had become disoriented by the smoke. Instead of turning left, where there was a nearby exit, he turned right and died near the exit to the Wilson Room, after being incapacitated.

The other people who died on the Commons had more time than the people in the rooms along the North Corridor did to move around before incapacitation. The pattern of fatalities and the time element in that pattern bears out the first witness's observation that the smoke and fire traveled faster down toward the Haight Room than in any other direction.

**Table 7.1.** Location of the Fatalities

| Fatalities | Location |
|---|---|
| 11 | Haight Room |
| 8 | North corridor between Haight, Disbrow B |
| | Commons: |
| 2 | Near Harrison Room |
| 3 | Near Wilson Room |
| 2 | Disbrow A |
| 26 | |

Of the twenty-four people injured, the great majority suffered from smoke inhalation. Some suffered only smoke inhalation, and others had that as well as other injuries. Of course, several of the people who raced the fire down the corridor suffered from burns. Those who exited early at the loading dock had reddened faces. As more exited, the reddening was lower and lower on the body, a change that showed that the layer of smoke and hot gases occupied more and more of the cross-section of the corridor and approached the floor.

People who were attending the seminar in the Harrison Room saved themselves by jumping out of the window. Besides smoke inhalation, their injuries included gashes, sprains, bruises, concussions, and broken bones. Some of them fell on the broken glass from the window and lay dazed among the shards and splinters.

The ratio of dead to injured was quite high—there were more dead than injured (26 dead, 24 injured). Nationally, for all fires, one person dies for every twenty-five persons injured. The Stouffer's Inn Fire exemplifies a characteristic of large, synthetic-fueled fires in public places. The dead-to-injured ratio was very high, as was the proportion of those present suffering death or injury. Significant involvement with the smoke takes a high toll.

The fire damage was strangely spotty. Although the carpeting was burned thoroughly in the path of the fire, in spots where it was covered by something, it was undamaged. The coat rack, furniture, and bodies of fatality victims left islands of carpet in a sea of ash. Yet, we know that this fire was hot, over 1,000° C. Firefighters were forced back by the intensity of the heat. The heat diffused through the ceiling and ignited the electrical insulation, which burns at 600°C. The heat beneath the ceiling must have exceeded 1,000°C. Yet, the bodies in the path of the fire weren't completely burned. Areas of carpeting protected by furniture and bodies didn't burn. How could this happen in such intense heat? The answer appears to be in the speed of the fire front. The rapid and directional movement confined the very high temperature zone of the fire to a narrow band that could not touch any fuel for a long enough time to consume it completely. For another example of a rapidly moving, superficially burning fire, see Chapter 6, which describes the MGM Grand Hotel Fire. Fuels that rapidly release combustible gases at low temperatures particularly feed fires of this type.

Because of the rapidity of the fire spread, firefighters required forty-five minutes after arrival to control the blaze. The injuries and deaths occurred before firefighter arrival, but much of the damage to areas off the involved corridors took place during the time the fire was being fought. Those who saw the windows in the Commons blow out also testified that the flames coursed through that area rapidly and then died down rapidly, confirming the theory that the fire had a rapidly moving front without much of a "back," at least in the corridors.

The wallcovering also burned along the corridors. In most of rooms, it decomposed just under the ceiling. This decomposition accounted for much of the heat damage from the fire in rooms that had no direct fire damage. The fire entered Disbrow A and B and the Receiving and Holding Room. Heat damage affected the Harrison Room and the Jamison Room. The Haight Room had some direct fire damage—mainly heat and smoke. Most areas of the floor were coated with smoke. A lieutenant from the White Plains Fire Department has a color photo taken in the kitchen right after the fire. The picture is of lettuce and tomato salads that were sitting in the kitchen, and the photo looks like a black-and-white shot. The lettuce and tomatoes look gray, in gray bowls. The uniform soot coating that was on all surfaces deprived all the substances of their normal colors and imparted to them the image of death.

About twenty to thirty minutes into the fire fight, men working on the roof were shaken by an explosion that was a rumbling quake similar to the one that occurred in the Younkers Brothers Department Store Fire. As in that explosion, parts of the building separated from each other: the raised roof area above the Grand Ballroom accumulated products of combustion and separated from the building when these products exploded. At this point in the fire, large quantities of fuels had been decomposed and could have generated large quantities of hydrocarbons. Nearly everything present, except the piano, was plastic: PVC and urethane chairs, polyester drapes and tablecloths, Naugahyde vinyl (PVC-ABS-modacrylic blend) and urethane foam sofas, and flooring of various kinds (nylon/wool, nylon, and PVC tiles). All of these materials decompose to release large quantities of hydrocarbons.

## THE FUELS OF THE FIRE

The description of the course of the fire, the locations of the fatalities, and the patterns of fire damage show that the fuels in the early stages largely determined the outcome of the fire. The fuels were the carpeting and wallcovering along the corridors. Although a wide variety of materials eventually turned to smoke and ash, both the NFPA and the experts for plaintiff's counsel zeroed in on the carpeting and wallcovering.

In fact, there is nothing unusual about this carpeting and wallcovering. Materials very similar to them can be found along the corridors of most of the new generation of pseudo-luxurious hotels and motels, which is why these mass fatal fires are so frequent.

The carpeting met all of the applicable standards for ignitability and flame spread. But when it was tested for fume and smoke chemicals, it emitted not only hydrogen cyanide, but also nitrogen dioxide, a potent pulmonary irritant that turns to nitric acid in tissue. Of course, the usual carbon monoxide was present also. The attitude of the NFPA and governmental investigators was one of puzzlement. This carpet was not supposed to burn like that! But obviously it did. Researchers at Factory Mutual Labs, an insurance industry facility, have known for years that materials in well-ventilated fires will burn when they may not under the standard test conditions.

Both the NFPA and counsel's experts found that the vinyl wallcovering would emit large quantities of decomposition products when subjected to unusual heat. Bubbles would appear under the surface of the plastic, eventually burst the surface, and release gases. Those gases would flare with an intense flame. This flame obviously contributed to the rapid, under-the-ceiling spread of the fire along the corridors. In addition, counsel's experts found that the gases included high levels of the acid gas hydrogen chloride, and phthalates, which are quite combustible.

Combined, these ordinary carpets and wallcoverings are deadly.

In the corridor were two or three sofas covered with Naugahyde vinyl, (which is a mix of PVC, ABS, and mod-acrylic), and stuffed with urethane foam. The artificial Christmas tree was PVC. A coat rack full of coats was near the Haight Room, in the path of the fire. The materials of the coats were not determined. All of these pieces of furniture and clothing added to the fuel load, and certainly the sofas and Christmas tree added their acidity to that of the wallcovering.

In most of the rooms off the burning corridors, the wallcovering decomposed, at least near the ceiling. It is possible that some of the toxic effects experienced in the rooms themselves derived both from the smoke and fumes entering the rooms from the corridors and from the acid gas generated in the rooms from the decomposing thin film of PVC wallcovering. It is impossible, of course, to parcel out which product was producing what toxic effect in those first few minutes of the fire, when life and death were fixed for the occupants of the third floor.

The chairs and other combustibles in the rooms that burned may have influenced the length of time the victims lived. The PVC covering on the chairs and the carbon monoxide generated by the polyester fabric and the acrylic certainly contributed to rapid death. Whether additional survivors could have been rescued by the firefighters if these furnishings and fabrics weren't present is something we will probably never know. We do know that the carbon monoxide and cyanide levels in the blood are roughly inversely proportional to the ages of the victims and may indicate how long the victims breathed in the fire atmosphere before death.

In the fire, many other fuels eventually burned. Cumulatively, these fuels contributed to the damage from smoke, heat, charring, and the explosion. Draperies, decorations, tables, the electric insulation, even the mastic holding the ceiling tiles onto the ceiling became fuels. They could only contribute to the property damage, and in this fire these fuels had little influence on deaths and injuries.

The three major factors that determined this fire's outcomes were the two initial decomposing materials (the carpeting and wallcovering), the design of the building and location of the primary fire, and the fire safety systems and procedures followed once the fire was discovered.

The people in the rooms along the North Corridor and in the corridor itself had no way to exit the building except from the loading dock. Because the fire was in the corridor itself, survivors raced smoke and wallcovering-spread flames down the North Corridor. Delay in reaching the decision to run that way or to jump out of a window meant death. Even making the decision in time left many people with respiratory and other injuries that are, most likely, permanent.

There was no exit from the Commons except at its southern end. (See Figure 7.1.) Failure to rush to that end meant death.

No one heard any fire alarms. This was because the fire alarms had been turned off manually, according to a hotel policy that was put into practice to reduce false alarms. The only sprinklers on the third floor activated long after people began fleeing. The architect had placed them only at the south end of the Commons. (According to the Harrison building code of that time, the hotel didn't have to have any sprinklers at all!) People exiting from the Wilson Room, the last people on the third floor to survive, got a little shower. The first people to leave that end of the Commons came out dry.

The air handling system operated during most of the fire, and had only a few operating smoke detectors in the ducts and no automatic system of turn-off connected with the smoke detectors. It continued to operate and help circulate the smoke.

The hotel staff was untrained in emergency procedures of this nature and some of them ended up in the hospital with smoke inhalation injuries. Indeed, the telephone operator notified the fire department only after being told by a smoke-sick employee that the alarm was real.

At this stage of its existence, the Westchester Stouffer's Inn main building was a big fire waiting to happen. Between the negligence of the town code, the hotel management, and the construction company and its suppliers, this fire was inevitable.

The combination of the carpeting and wallcovering with the corridor design created a fire that moved faster than most of us could envision. Of the many witnesses who testified in the case and tried to describe the speed of the movement, one in particular conveyed the picture.

> He stepped through the Haight hallway door and looked out and up. He saw a river of fire travelling at the speed of a freight train overhead beneath the ceiling and judged it to have been about two feet of pure flame and dense black smoke.... They came upon the doorway to the receiving area. As the flames raced down the hallway and approached the sill, it was like a liquid fire. It came down and went back up.... The flames were going much faster than he could go.

## THE VICTIMS

Table 7.2 lists the observations that were grouped from the autopsy protocols. These observations involve soot deposition or unusual internal organ appearances. Many of the organs indicated the presence of highly irritating soot. Some of the symptoms that indicated this were lung edema (fluid),

**Table 7.2.** Autopsy Observation Prevalences.

| Observation | Number of Victims | Percentage of Symptoms |
|---|---|---|
| Lung edema | 26 | 100% |
| Airspace destroyed | 25 | 96% |
| Upper respiratory edema | 16 | 62% |
| Goblet cells increased | 19 | 73% |
| Soot throughout respiratory tract | 26 | 100% |
| Cardiac edema | 8 | 31% |
| Congestion or engorged vessels in organs | 26 | 100% |
| *Kidney* | 13 | 50% |
| *Brain* | 4 | 15% |
| *Liver* | 10 | 38% |
| Spleen | 4 | 15.7 |
| Soot in digestive tract | 10 | 38% |
| Internal hemorrhaging | 15 | 58% |
| Eosinophilia | 11 | 42% |
| Soot in nostrils/mouth | 9 | 35% |

destruction of airspace, upper respiratory tract edema, goblet cell increase, and eosinophilia in the respiratory tract. **Goblet cells** are mucus cells and protect the respiratory tract against irritation. **Eosinophilia** means that a certain type of white blood cell is present in high numbers that characteristically appear when irritating foreign matter is introduced into the body. These observations differ greatly from those of simple carbon monoxide deaths from the burning of traditional materials.

Tables 7.3 and 7.4 show the number of victims with various levels of carbon monoxide and cyanide. Over half of the victims had sublethal levels of carbon monoxide in their blood. One victim had very low carbon monoxide (28 percent carboxyhemoglobin) and a low cyanide level (0.2 microgram per millileter). Something else in the smoke besides

**Table 7.3.** Levels of Carbon Monoxide in the Blood of Fatality Victims.

| Range of Carboxyhemoglobin (COHb) | Number of Victims in That Range | Percent of Victims |
|---|---|---|
| 16–20% | 1 | 3.8% |
| 21–25% | 0 | 0 |
| 26–30% | 5 | 19.2% |
| 31–35% | 4 | 15.4% |
| 36–40% | 4 | 15.4% |
| 41–45% | 2 | 7.7% |
| 46–50% | 3 | 11.5% |
| 51–55% | 1 | 3.8% |
| 56–60% | 2 | 7.7% |
| 61–65% | 1 | 3.8% |
| 66–70% | 3 | 11.5% |

Carbon monoxide average: 42.69%, standard deviation = 14.17%
Fifty-four percent of the victims had COHb levels below 40 percent.

**Table 7.4.** Levels of Cyanide in the Blood of Fatality Victims.

| Range of Cyanide (CN) | Number of Victims in That Range | Percent of Victims |
|---|---|---|
| 0 | 1 | 3.8% |
| 0.1–0.5 | 2 | 7.7% |
| 0.6–1.0 | 3 | 11.5% |
| 1.1–1.5 | 12 | 46.2% |
| 1.6–2.0 | 3 | 11.5% |
| 2.1–2.5 | 1 | 3.8% |
| 2.6–3.0 | 2 | 7.7% |
| 3.1–3.5 | 2 | 7.7% |

Cyanide average: 1.47 micrograms/millileter, standard deviation = 0.83.
Twenty-three percent of the victims had CN levels below 0.6.

carbon monoxide and cyanide was contributing to the deaths. Something in the smoke also caused rapid incapacitation, so that people seemed to drop where they were when they came into contact with the smoke. This something is corrosive acid gas.

## INJURIES

Out of the twenty-four people injured, eleven had their hospital records sent to plaintiff's lead counsel for review by the team of experts. A brief summary of their injuries as a group is listed in Table 7.5

The respiratory symptoms and the abnormal blood pH and blood gases seen in Table 7.5 are typical of inhalation of acid gases. The respiratory tract is injured by the acid and the body tries to compensate for the intake of acid by what is

**Table 7.5.** Injuries of Eleven Survivors.

| Number Out of Eleven Survivors | Injury/Symptom |
|---|---|
| 10 | Smoke inhalation noted on admission |
| 8 | Abnormal blood pH and blood gases |
| 8 | Traumatic injuries (two fractures, all eight with cuts, bruises, sprains) |
| 6 | High white blood cell counts |
| 6 | Respiratory symptoms (lung sounds, shortness of breath, wheezing, coughing, hoarseness, coughing up sputum) |
| 5 | Acidic urine |
| 4 | Fever |
| 2 each | Gastritis, burns, kerato-conjunctivitis (inflammation of the cornea) |

called respiratory compensation. The blood pH becomes alkaline when this overcompensation occurs, and blood concentrations of carbon dioxide and bicarbonate become too low, which accounts for the abnormal blood gases. Oxygen concentration is also low because of the respiratory tissue damage.

The respiratory symptoms show that the whole respiratory tract could be injured in this type of smoke, from the deep lungs to the upper tract where the vocal cords sit. Four people experienced fever, and several had a high white blood cell count. These symptoms are typical of fire survivors with inhalation injuries. The immune system is trying to clear the lungs of particles by having the white blood cells engulf and remove them. These particles are toxic to white blood cells. Fever often arises in cases of what is called granuloma disease (large numbers of granular white blood cells). There have been articles in several medical journals featuring fever following exposure to burning plastics. When accompanied by flu-like symptoms such as coughing, shortness of breath, and chills, it is called polymer fume fever.

PVC is well-known as a source of the injuries listed in Table 7.5. Indeed, the one party who forwarded his continuing medical records to lead counsel showed a typical delayed or continuing reaction to PVC smoke inhalation: sensitivity to dust and smoke, loss of lung elasticity (over-inflated lung), wheezing, and airways sensitization and nasal congestion. He also regularly suffered from sleep disruption.

## THE INVESTIGATION

Lab tests on the carpeting in the Stouffer's Inn showed that a number of chemicals were emitted during combustion. The tests showed that nitrogen dioxide, hydrogen cyanide, carbon monoxide, carbon dioxide, and hydrocarbons were emitted. Outside of the sofas in the Commons, the carpeting represented the only possible source of the cyanide found in the fatality victims' blood. Other materials that could have emitted hydrogen cyanide burned later in the fire, possibly after the victims died and the first bodies were found by the firefighters.

Lab tests also showed that the wallcovering was plasticized PVC and that it emitted large quantities of hydrogen chloride and phthalate. The National Bureau of Standards combustion toxicologists analyzed soot samples and found elements that were consistent with a mixed origin of carpeting and wallcovering.

The lines of evidence that identify the carpeting and wallcovering as the origins of the early killing smoke are:

- Eyewitness accounts
- Autopsy findings
- Injury patterns
- Lab tests on materials
- Soot analysis

During the 1974–1978 period, fire reconstructionists at both the NFPA and federal government "discovered" the phenomenon of the corridor fire. Fire in carpeting that is placed in a corridor will spread with great rapidity compared with fires in a square room. This is because airflow and the fuel lie essentially along an axis, not radially. Smoke and gases also spread more rapidly down a corridor. If the ceiling and/or wallcovering also burn or emit combustible gas, the effect is greatly heightened.

The Stouffer's Inn fire was a corridor fire of the worst kind. Yet, the NFPA mutes the significance of the corridor configuration. Perhaps this agency was already jaded by 1981. Indeed, the NFPA report plays up the presence of the coat rack in the corridor and concludes that this is what gave the fire its intensity and rapidity. Such a conclusion flies in the face of eyewitness accounts, the chemistry of the inhaled soot, autopsy protocols, physical evidence, analysis of combustion products of carpeting and wallcovering, and hospital records of casualties. The NFPA, with its tunnel vision, distorted reality by blaming the outcome of the fire totally on code violations. The fact that its own research team saw the wallcovering bubble and burst forth with combustible, flaring gases in lab tests meant nothing to the authors of the report.

The Stouffer's Inn Fire combined all the worst elements of a classical corridor fire: burning floor covering, rapidly decomposing wallcovering, a relatively low and heat-reflective ceiling, and a long, narrow corridor that channeled all forms of fire product in one direction. The occupants of the meeting rooms along the corridors, especially the north corridor, raced the smoke and wallcovering-generated flames for their lives. If the coat rack had *not* been there, they would have had to do the same thing.

## LESSONS OF THE STOUFFER'S INN FIRE

The origin of the Stouffer's Inn fire was attributed to arson. But for most of the fire reconstructionists and the corporation, the plaintiffs, and the firefighters, the issue of arson ranks low as a factor in this fire. It is likely that the cause of this fire will never be known for certain. It could have been arson or it could have been human error.

For too many fire departments and newspapers, the word "arson" means that nothing else about the fire outcome is worth analyzing. The word "arson" is supposed to explain away everything, when, in fact, it explains very little. The fire had a lag phase, like all fires. It didn't become a racing, roaring inferno immediately. Arson may or may not have been responsible for the fire's origin. Once beyond the lag phase, many other factors interacted with the fire to lead to the deaths, injuries, and damage.

In many of these fires, the building owners and operators believe that if they raise the issue of arson, even on the flimsiest evidence (evidence in the Stouffer's Inn fire, while not as flimsy as usual, was not beyond reasonable doubt), that everyone involved in the case will let them off the hook.

Because of the emotional power of the idea of arson, which is rooted in sexual feelings, the course of investigation can become bizarre if any of the parties to the fire reconstruction are vulnerable to that emotional power.

The next lesson from the fire emphasizes corporate responsibility. The fulfillment of the requirements of a local building and fire code does not necessarily mean that a corporation has fulfilled its responsibilities for the safety and health of its patrons and employees. This is because local codes frequently fall very short of reasonable standards. Many corporation think it's economically shrewd to build factories, hotels, residential housing, or offices specifically in municipalities or counties with grossly outdated, weak codes. Short-sighted, unethical strategies of this nature generally become compounded by such tactics such as the silencing of the fire alarms by the switchboard operator in the Stouffer's Inn.

The laws that parcel out liability for chemical landfills can also apply to the case of toxic exposure from fires. Everyone along the chain of responsibility bears liability: the owner and operator of the corporation, the suppliers and distributors of these plastics, the original manufacturers who sold the material to the distributors, and the public and private agencies (local government and insurers who oversee these matters.) This view of liability has been borne out in litigation, including the settlement of the multitude of suits that resulted from the Stouffer's Inn Fire. Subrogation suits abounded.

Thus, the lesson on corporate responsibility is briefly, "let the buyer and the seller beware."

The third lesson teaches us what most of these large fires show—namely, that the people were killed and injured by the smoke, not the flames. The incredible speed with which the smoke moved in a wave was more of a factor in death and injury than the unusual rapidity of the flame spread, although the two obviously went together. Smoke entered rooms that had little or no fire damage and caused deaths and injuries there. Several of the fatality victims had very low levels of carboxyhemoglobin in their blood and for these victims, at least, carbon monoxide was not a factor in death. One victim had low carboxyhemoglobin and low concentration of cyanide. The active poisons in the smoke included high concentrations of irritants that must have been the hydrogen chloride from the PVC wallcovering and may have included nitrogen oxides from the carpeting. Smoke inhalation hazards are quite different now from what they were before the widespread use of synthetics, when carbon monoxide was the main concern.

A fourth lesson from this fire that was not explicitly discussed by the National Fire Protection Association was the interaction of the placement of the particular fuels and the design of the building. The third floor consisted almost entirely of corridors and of the rooms on each side of them. Uninterrupted combustible floor covering and/or wallcovering along these corridors invited a classic corridor fire. Nothing in any of the building or fire codes discusses uninterrupted combustible finishing. The codes describe only finishing that meets certain performance standards determined by tests that have nothing to do with *real* fire conditions.

It was bad enough that there was uninterrupted combustible finishing in the corridors. But worse yet were the particular individual finishings and their relative placement. The floor finishing consisted of nylon/wool carpeting with jute padding, which ignites at a lower temperature than the PVC wallcovering, but PVC *decomposes* at and below the carpet ignition temperature. Thus, along the upper wall right below the ceiling where the radiant heat accumulated, the plasticized PVC rapidly unraveled chemically and released its acid gas and combustible plasticizer. The carpeting on the floor and the PVC on the wall formed a long rectangle of death and injury. The carpeting and the PVC both passed the performance standards of the NFPA and ASTM (American Society for Testing Materials).

The characteristics of corridor fires are massive heat accumulations at the ceiling and accelerated spread of heat, smoke, and fire from the directional channeling of air movement and energy. In such fires the performance standards of the NFPA and ASTM cannot protect life and health. Simply stated, *synthetics should not be used freely in corridors, especially corridors of institutions and public places.*

Finally, the autopsy results and hospital records indicate that the smoke intoxicants of this fire included other toxins besides the traditional carbon monoxide and cyanide detected in the blood of the fatality victims. The victims could not move very far; they succumbed rapidly. The irritants and narcotizing organic chemicals that are present in the smoke from synthetics are famous for their ability to rapidly incapacitate. The lung edema, hemorrhaging, and irritation found during autopsy underscore the presence of corrosive irritants in the smoke. The Stouffer's Inn fire forms another link in the chain of fatal fires in which fumes and smoke of nontraditional chemicals killed. A wrong turn was enough to fell one victim very rapidly, the last person out of the Wilson Room. *Smoke from synthetics is unlike smoke from natural materials.* This fact cannot be repeated enough.

*CHAPTER 8*

# PATH Subway Fire

PATH is the acronym of the Port Authority subway system, which links several New Jersey cities with each other and with New York. Every day, tens of thousands of commuters trust PATH to get them to their jobs and home again.

The riders who boarded the 7:30 a.m. train on March 16, 1982 were no exception. The train from Hoboken to New York City began its trip right on time.

Shortly after this train entered the tunnel under the Hudson River and had nearly completed the brief trip from New Jersey to Manhattan, the electrical system started to malfunction. One thing after another became affected. Finally, the air brakes could not be released, and the train was stuck.

Soon after the airbrakes locked, acidic smoke coming from under the second car rose up and into the car and began moving through the train. This cloud of smoke was only the beginning of an hour-long ordeal in that crowded set of subway cars which left sixty people in need of medical attention, and twenty-five of them needing hospitalization.

All of the injured people (commuters, trainmen, police officers) suffered from smoke inhalation.

The tunnel was dark and acted like a horizontal chimney. Before the PATH management finally stopped all traffic in both tubes of that tunnel, the piston action of moving trains drove the smoke back and forth onto the trapped people.

The hundreds of people on that ill-fated train finally reached fresh air and light by walking the track and climbing a narrow emergency stair to Manhattan. Because of the narrowness of the stairs, only a single file could climb at a time. When firefighters tried to get down to the track to rescue people still trapped in the back of the train and to put out the fire, exiting commuters blocked the stairs. Firefighters had to wait until the commuters had ascended before they could perform rescue and extinguishment.

At the time of the fire, PATH had already formed a special committee to explore the fire safety needs of the subways and to develop recommendations. But the committee was frustrated by the time of this fire, because it was having difficulty implementing its suggestions.

The PATH system runs light rail trains between Manhattan and several New Jersey cities situated on the banks of the Hudson River: Jersey City, Hoboken, Newark, Harrison, and others. These cities could almost be considered part of New York because of the connections via the PATH system, the Port Authority bridges and tunnels, and the three major airports.

During the average workday, the PATH system transports 60,000 commuters. In addition, the Holland and Lincoln Tunnels facilitate thousands of trips by cars and buses. A veritable human termite nest buzzes with activity under the Hudson River. The design of these tunnels is old and was drawn according to the conventional wisdom of tunnel engineers of the period. Ventilation, emergency exits, communications, and surveillance for unusual events are all heavily determined by the basic conventional tunnel design.

## THE FIRE

The train left Hoboken and soon developed electrical problems in the eastbound tube of Tunnel B within the PATH system. Hundreds of people were trapped under the Hudson River in the tube. It was rush hour, and the PATH management does not turn off the third (electrified) rail during rush hour. Hundreds of people were trapped in this horizontal chimney, with airflows pushed around by the train traffic. But PATH management will not free a tunnel from traffic during rush hour. The trains must keep running. Trainmen screamed over the radio about the black, choking smoke and the fainting passengers. But PATH management knows little about the plastics in its own cars' wire insulation and cable sheathing and the toxic fumes that these plastics emit.

A court case resulted from this fire and papers filed by PATH in this case showed that the management and its consultants believed that if they proved the wiring had no PVC associated with it, then they were blameless with respect to materials. They proudly presented backup data documenting the wire insulation as hypalon. But hypalon is chlorosulfonated polyethylene. It is toxic. When it burns, it emits (besides carbon monoxide) chlorosulfonic acid, hydrogen chloride, and sulfur oxides.

At 7:12 a.m. the conductor and the motorman of the Manhattan-bound train talked about "loss of indication." "Loss of indication" means that the signals go dead that show whether doors are opened or closed. After some flipping of switches and more conversation, they decided the problem was solved, and the train began to move.

The train stopped again when electricity was lost from the third rail. There was more conversation and more fiddling with switches. Then the motorman thought he had solved the problem and said that he'd move the train—but he could not. Enough air pressure could not be built to release the pneumatic brakes, so the train sat under the Hudson.

Then word came to the conductor and the motorman that the chassis of the second car was smoldering and the smoke was entering the passenger compartment. The time was now 7:19. The second car was evacuated. First the passengers were told to move forward. Later, they were made to recross the fire car and move to the rear of the train.

A railroad motorman on his way to work was in the rear and worked the radio. He began a nearly hour-long conversation with the dispatcher and managers. He described the smoky condition and the increasing panic among the passengers. Some of the passengers fainted, and others cried, screamed, and trembled during the ensuing ordeal. Everyone was sure that death was near. The smoke burned their lungs and smelled deadly. One of the passengers was a young woman in her seventh month of pregnancy. Another young woman stayed with her and tried to help her through the episode.

The radio conversation that was transmitted back and forth between the PATH dispatcher's office and the train clearly showed that there was no comprehension in that office of the conditions in the tunnel. The fire wasn't reported to the Fire Department until 7:27, a delay of ten to twenty minutes. Approximately four minutes were taken up because the call to the New York Fire Department was repeatedly interrupted by other business! The call began at 7:23, and the necessary information was not conveyed until 7:27. The man who delayed the reporting was later rebuked, but most of the fault lies with the PATH management. The management failed to do anything until conditions had deteriorated to the point of grave endangerment to everyone on the train. This happened because they wanted to keep the system running at all cost. They literally refused to accept the fact that a horror was happening.

Finally, after many minutes of conversation back and forth, PATH management agreed to turn off the electrified third rail so that the passengers could walk alongside the tracks to the Morton Street emergency exit. By this time, the fire in the second car had grown to such proportions that the passengers who had been herded to the back of the train could not escape without help. The passengers in the front simply walked to the exit, although they too became repeatedly exposed to smoke because of the piston action of the trains in the tunnel. The darkness made them stumble and grope for their way.

The passengers waited in the rear of the train. There was more conversation back and forth. At 7:25 PATH management decided to keep trains out of the tunnel because motormen and conductors on the other trains complained about the heavy smoke: "We have a heavy smoke condition Tunnel A midriver." Pleas from the off-duty motorman who was trapped with the passengers in the rear of the fire train described the smoke, the darkness, the panic, and the deteriorating physical condition of the crowded passengers.

PATH police came from the New Jersey side, equipped with SCBA (self-contained breathing apparatus), and soon afterward NYC firefighters, also in SCBA, reached the track by swarming down the narrow stairs against a tide of escaping passengers. Police and firefighters led trapped passengers onto the track and to the stairs that led up to the street. One at a time, they climbed those stairs to air and light. They coughed and gasped, coated with acidic, greasy soot, but they got out from the dark, the closeness of the tunnel, and the smoke.

Vans and ambulances vied with firefighting rigs for space on Morton Street. The block was filled with red trucks, white trucks, and the white-and-orange of the public hospital ambulances. The white coats of the private hospital paramedics, green uniforms of the public paramedics, and black-and-yellow turnout coats of the firefighters swarmed near the emergency stair exit, a welcome but disorienting sight to the dizzy, already smoke-disoriented passengers.

## THE CAUSE

The fire had taken place in the motor control area of the car chassis, an area known as the group switchbox. Although from its name it sounds like a little electronic gizmo, the group switchbox measures about five by four by three feet and is full of wires, wire insulation, and cable sheathing around a steel support structure. An unfortunate design flaw of these cars places this heat-and fire-prone switchbox near the pneumatic tube of the air brake. This tube, generally, also is made of plastic, usually polyethylene. The plastic pneumatic tubes in the subway cars of the New York City system caught fire repeatedly.

The power wire insulation in the group switchbox was made of hypalon (cross-linked chlorosulfonated polyethylene). Smoke from this plastic would closely resemble smoke from PVC in its physiological effects—it is acrid, dense, and greasy, with a high proportion of the particles in the respirable and lung-retentive size range. The major illnesses and injuries from inhaling the smoke would resemble those of PVC smoke: loss of immunity to respiratory tract infections, sensitization of airways (maybe even true asthma), lung damage, phlegm production, sinus trouble, problems with peripheral microcirculation, heart problems, and neurological/psychological disorders such as those described in previous chapters.

The other plastics present in the cable sheathing were a nitrile (probably acylonitrile) and PVC. These would also produce irritants. The nitrile produces cyanide in a fire. However, the major fuel by weight, and the earliest fuel, was the wire insulation. It would produce chlorosulfonic acid, hydrogen chloride, sulfur oxides, and hydrocarbons, as well as carbon monoxide and carbon dioxide.

The decomposition and combustion of these irritant-producing plastics didn't take place in an above-ground building, as in the other fires we have discussed, but in a tunnel. Even with "normal" controlled combustion, tunnels cause health-threatening air quality problems because of their ventilation patterns. An example of this kind of problem is the air pollution levels attained in the Lincoln and Holland Tunnels, which are also owned and operated by the Port Authority. At times of heavy traffic clogging, drivers and passengers have to be pulled from the cars and buses and given bottled oxygen.

Tunnels are horizontal chimneys and, simply by their tubular shape, cause wind. When trains or vehicles travel through tunnels, their motion acts as a piston movement: it heightens the flow of air and air pollutants through the tunnel. If trains move, they also cause smoke movement if a fire breaks out in the tunnel. Thus, even after leaving the burning train and walking along the track, the passengers received doses of smoke, and passengers and PATH employees on other trains also received a dose. There is no way smoke can diffuse out of a tunnel. Tunnel ventilation patterns also cause the heaviest concentration of pollutants closest to the tracks. The passengers couldn't win.

The passengers had two built-in strikes against them: the plastic wire insulation and cable sheathing, and the fact that the event took place in the tunnel. What more could go wrong? What element was left either to improve or to degrade their plight? The answer: the PATH managers and their policies. Like most transportation systems, whether rail transit or bridge-and-tunnel systems for cars and trucks, PATH has policies that encourage keeping the system working—even during some potentially hazardous situations. Because of this policy, early intervention into worsening conditions simply doesn't happen. Should an incident occur during rush hour, the trains will continue to run.

Humans don't like to face unpleasant realities. We like to deny that horrible things are happening. Most of the so-called "human errors" that degraded the already severe ordeal for the passengers sprang from PATH management's denial of the seriousness of the fire. The motorman pleaded for help and described the situation in detail to the managers at the other end of the communication system. The managers said, "Sure, sure," and turned their attention not to the fire but to minimizing the effect of the fire on rush hour traffic. Failure to call the Fire Department early, which PATH admitted, was only part of a larger pattern.

## THE VICTIMS

My file contains the medical records of only two out of the sixty or so people who received medical attention as a result of the PATH fire. Partial data, or even a very small percentage of the possible data base, are the frequent handicap under which an expert consultant must work. Opinions must be rendered on the basis of the preponderance of the evidence, as in a regular civil suit. Each scientist-consultant must decide how much evidence is too little evidence for rendering an opinion. In some states the court asks for a "scientific certainty" as a basis for expert witness opinion. In the context of today's bitter debates over the meaning of statistical tests, which statistics should be displayed, and "significance levels," the phrase "scientific certainty" betrays the trust of the lay community toward the folks in the white coats with the computers. We in the white coats, meanwhile, struggle to educate others that mathematics isn't science, nor is the field of statistics. Science uses statistics and mathematics as tools, and other tools are part of the total kit. "Scientific certainty," indeed!

The immediate symptoms of the two young women in my file show that both had the immediate symptoms that afflict survivors of smoke inhalation from halogenated plastics-fueled fires. One woman complained of dizziness, headache, a burning respiratory tract, and increasing difficulty in breathing. Her blood gases changed drastically during her first twenty-four hours in the hospital, as did her white blood cell count. She produced black phlegm. She had tachycardia as well. On top of this, she was seven months pregnant and obviously quite anxious about the baby.

The other woman had a similar set of immediate symptoms: headache, burning respiratory tract, coughing and wheezing, dizziness and confusion, and abdominal pain. Her hospital records show that she had difficulty breathing, headache, phlegm production, and gallstones. Only the gallstones were not connected with the smoke inhalation. Her blood gases also showed changes similar to those of the first woman. She registered extreme anxiety, fearfulness, and sadness during her hospital stay. Her respiratory symptoms did not entirely vanish by discharge time.

Each of the young women who required medical attention after the evacuation sustained respiratory injuries that persisted at least five years after the fire. In addition, one woman now has a heart murmur and the other, possible exertional tachycardia. They show different injuries to the same organs. The lung malfunctions show up as asthma in one woman and as "overinflated" lung in the other, i.e. development of a deadspace in the lung due to loss of elasticity with compensation for the dead space by increase in lung capacity.

The woman with "overinflated" lung and possible tachycardia has other physiological problems: calcified nodules in the lung, development of calcified nodules in hands and feet, and weight control difficulty because she cannot exercise. Although the baby she was carrying at the time of the fire has proven normal and healthy, the fears for herself and the baby left her with lasting psychological scars. Her relationships with her husband and others close to her changed greatly. She was divorced.

The woman suffering from asthma and heart murmur has other physiological problems as well. They include headaches, phlegm production, sinus trouble, and extreme sensitivity to air pollutants, especially dust and smoke. She also has heavy, painful menstrual periods. The experience left her psychologically scarred, with depression, anxiety, nightmares, and sleep problems, and the inability to concentrate much of the time.

Both of these women have required regular medical attention for monitoring their respiratory and heart conditions and for treatment. They have both had extensive psychotherapy.

The prognosis for these women is that, at best, they can expect to retain their present level of physical problems and, at worst, to suffer some deterioration. Hearts and lungs are notorious for insidious and progressive deterioration, once they are impaired by chemical lesions.

## WHOSE FAULT WAS IT?

When PATH was faced with the prospect of going to court, their attorney reacted by putting off the court date at least eight times, with a different excuse every time. They did not hire a technical expert until the seventh attempt to bring the case to trial, and this expert was only a transit electrical maven, not a materials engineer, polymer scientist, or fire safety expert. The PATH attorney begged to put off the court date yet another time because they had not retained a medical expert! The management style of PATH obviously derives from the philosophy that if you ignore a problem long enough and procrastinate long enough, it will go away. As they procrastinated with the court matters, so they procrastinated in dealing both with fire prevention and protection measures before the fire and with the fire itself. Indeed, PATH is *still* procrastinating with fire protection; they have not yet upgraded the Morton Avenue emergency exit or the emergency ventilation system for the tunnel, according to a *New York Daily News* article of May 26, 1987. Now *that's* procrastination!

The management style of PATH should be classed as clever-stupid. They thought that if they could prove that the wiring is not PVC-insulated, they would be in the clear. So the report written by their technical expert reads like a strange litany with the recurring chant "this was not PVC." They offered no written pretrial testimony on the fire behavior of the wire insulation or on the chemistry of the smoke. They offered no evidence on the toxicity of this smoke or on the effects of tunnel ventilation on the concentration of the smoke.

The technical isolation of this management causes (or should cause) astonishment. This system transports tens of thousands of people each way every workday! Yet, the management did not know how to pull together a proper team of experts either to perform its own investigation of the fire or to testify in court to give PATH'S side of the story. Their investigation was spotty, overdeveloped in metallurgy, and underdeveloped in polymer science, fire dynamics, and toxicology. Any consideration of toxicology was entirely lacking.

We should not be so hard on PATH management alone, however. Most rail systems in this country suffer a similar management style and philosophy. Europeans have tunnels with fire detectors and sprinklers. We do not.

## LESSONS OF THE PATH FIRE

### Lesson One

Although PVC most commonly causes severe long-term health problems, other plastics containing halogens can also cause similar severe chronic diseases. Hypalon certainly falls into this category of PVC-like plastics. Hypalon may be even less stable than PVC because the specifications guarantee physical stability at 90°C when dry and only 75°C when wet. PVC is guaranteed stable at 90°C wet or dry. Hypalon, by formula, would deliver the same weight of chloride to the air when decomposed as the same volume of PVC. Just because a plastic isn't PVC doesn't mean it's safer.

### Lesson Two

Acid smoke generation in an underground tunnel turns that tunnel into a horizontal MGM Grand Hotel. The tunnel acts like a sideways chimney. The smoke cannot diffuse and remains relatively concentrated. Movements of other trains repeatedly re-expose the passengers on the fire train to the smoke. The characteristics of tunnel ventilation increase the duration and concentration of passenger smoke exposure.

### Lesson Three

Many transit system managements simply do not have the philosophy or the technical understanding to cope with plastics and their intricacies. They rely on the literature of the vendors and are caught without proper fire prevention, passenger protection, and fire extinguishment equipment and contingency plans.

**Lesson Four**

Public agencies, in dealing with plastics-related emergencies and their consequences, may behave more callously and incompetently than even a private owner/operator of a commercial building. Because many governmental entities enjoy immunity to general negligence and liability laws, they do not feel that they have to be completely careful when purchasing materials, nor do they feel accountable for their failure to purchase proper materials.

The ordeal of the passengers on the PATH train had the same generic roots as the ordeal of the MGM Grand Hotel patrons, the conference participants in the Stouffer's Inn Conference Center, and the firefighters at the New York Telephone Fire. These roots are the inherently unstable nature of the plastics exposed to heat, the toxicity and corrosive irritancy of the chemicals released from the decomposing/burning plastics, and the design of the structure within which the fire took place. Regulations which may in the future be promulgated to minimize risk to life and health from plastics in buildings should also cover public and commercial passenger transportation such as trains, buses, and airplanes. The U.S. Department of Transportation and the National Transportation Safety Board have issued numerous analyses of individual fires on planes and trains, and reports on the fire safety of materials used on planes and trains. A continuing philosophy of non-regulation or deregulation would mean that the present clearly unacceptable, well-documented risks will remain and that many more will die. What is needed now is regulation and the support of concerned citizens.

*Chapter 9*

# The Watchdogs Sleep

A dozen or more governmental agencies and quasi-governmental private code-making bodies are supposed to protect the public from health- and life-threatening aspects of plastics. These governmental agencies can be found at all levels: federal, state, county, and municipal. The code-making bodies may be national or regional.

## FEDERAL AGENCIES

At the federal level, many agencies have a specific role in ensuring health and safety with respect to plastics in fires and other high temperature situations. These are the Consumer Product Safety Commission (CPSC), Occupational Safety and Health Administration (OSHA), Environmental Protection Agency (EPA), Federal Trade Commission (FTC), National Bureau of Standards (NBS), Health and Human Services (HHS), and Housing and Urban Development (HUD).

Although each of these entities has a different focus, there is a good deal of overlap. The Federal Trade Commission (FTC) and Consumer Product Safety Commission have some overlap in the area of labeling of products that pose potential toxic hazards in fires. The EPA has jurisdiction over indoor air pollution, but OSHA has jurisdiction over the air that workers breathe and over worker fire safety. The NBS, which is within the Department of Commerce, and the National Institute of Occupational Safety and Health (NIOSH) within HHS, both have primary health research missions. HUD overlaps somewhat with private code-making bodies because it traditionally has a required code for federally subsidized housing.

With all of this overlapping and so many agencies involved, we would expect good, tough regulation. We would expect competition to ensure health and safety. But that isn't what we're getting. These agencies have become "plasticized"—they have been influenced by the plastics industry to lessen regulations.

## THE URETHANE FOAM SETTLEMENT

As an example of how governmental agencies become plasticized, let us consider the 1975 landmark settlement between the Federal Trade Commission and those in the plastics industry who manufacture, use, and distribute flexible urethane foam. For over a decade this settlement has been considered a watershed, a stroke against threats to public safety, and a precedent-making outcome. As with many acts and objects surrounding or growing out of the plastics industry, this is not what it is purported to be.

In the 1960s and early 1970s, the manufacturers and distributors of urethane foam mattresses and upholstered furniture advertised that the foam was fireproof and self-extinguishing. The Society of Plastics Industries also distributed literature and helped with this marketing effort. Sales people were trained to use this claim when selling this product, emphasizing this as a positive attribute. These mattresses and pieces of furniture were much cheaper than their counterparts, which were stuffed with natural products such as cotton batting, animal hair, or feathers.

Unfortunately, there were a number of fatal fires that clearly showed that urethane foam was quite flammable and far from self-extinguishing. The fact that the main cause of death in these fires was smoke inhalation, compared with victims of fires caused by natural materials, shows that urethane foam had unusually toxic smoke as well. These fatal fires resulted in court cases against the manufacturers, distributors, and the SPI.

The Federal Trade Commission took note of these court cases and began its own investigation of the advertisements and the behavior of urethane foam. As a result of the investigation, the FTC documented the misleading nature of the advertising, the complete ignorance of the distributors and individual sales people about the behavior of the material in fires, and the flammable and toxic hazards associated with the material in fires.

The Commission filed suit in court against both the individual companies involved in manufacturing the foam and/or selling it to the public and the umbrella trade association, the Society of Plastics Industries. Lengthy negotiations took place between the complainant and the defendants. Shortly before the case was due to go to trial, the defiant stance of the defendants softened. The parties came to a settlement in a detailed agreement.

## The Settlement Agreement

The settlement agreement contained four key sections that were designed to protect the public from the fire hazards of urethane foam.

1. Advertising—There must be no more false advertising claiming that the foam is fireproof or self-extinguishing.

1. Labeling—A warning label describing the fire hazards must be affixed on products containing urethane foam.
2. Education of sales persons—Sales persons must be given instruction about the fire hazards of urethane foam and are expected to answer consumers' questions about them.
3. Research—A $5 million research program was required to explore the fire hazard of materials, techniques of fire retardation, and toxicity of materials in fires.

The settlement had no provisions for regulating the sale or use of urethane foam products. Thus, the fatal fires continue. As of 1982 urethane foam was estimated to be connected with approximately 20–35 percent of all fatal fires! These are the fires in mattresses and upholstered furniture. Now, because the FTC does not regulate the use of urethane foam or its sales, the same old issue of urethane foam-fueled fatal fires confronts such agencies as the Consumer Product Safety Commission. In order to carry out the settlement agreement, the plastics industry formed a special group, the Urethane Safety Group. The blatantly false advertising ceased, products were labeled, and sales people were educated. The research program was also set up.

However, the Urethane Safety Group went beyond fulfilling the settlement agreement: they instituted damage control. They set out to "rehabilitate" urethane foam in the eyes of the public. Their success in doing so without actually changing the hazards made the tactics of this group a model for other manufacturers of plastics. Many of the techniques of manipulation, obfuscation, and confusion pioneered by the Urethane Safety Group were subsequently used by the Vinyl Group and the Plastic Pipe Institute, among others.

These groups employed techniques of marketing around the hazards without blatantly lying. Misinformation was liberally given out to both the public and governmental officials. For example, deliberate confusion was fostered about the dangers of the flammability and toxicity of urethane foam. The constant refrain was that the material was fire retarded and couldn't be dangerous. Another line that was intended to obscure the truth was the assertion that urethane foam is all around us, everywhere, and therefore can't be dangerous. Then the old line appeared that all organic materials burn and emit toxic carbon monoxide that is responsible for death, so stop picking on urethane foam. The group also tried to imply that the research they did had a direct influence on the safety of the products. But in fact, the research resulted in little or no product improvement.

When the Urethane Group disseminated its public relations statements about the findings of the research program, no mention was made that this program was formed because of a settlement agreement that stemmed from criminal actions on the part of the industry, actions that resulted in deaths. The industry made it seem as though the research project was initiated by the industry out the goodness of its heart. Toward the end of the research program, the group became more sophisticated and began to require research of biased design to show that the urethane was safe, or at least acceptably risky. In reviewing the experimental designs described in the published papers of this research program, a scientist can see where the researchers conducted these experiments in ways that were designed to bias their results. Fortunately, much of the research occurred before this sophistication. This initial research shows that the chemicals given off by the foam are different from those given off by wood and pose different and more serious risks than those given off by natural combustible materials.

## The Campaign for Fire Retardants

A similar public relations campaign was pushed for the addition of fire retardants to urethane foam. In 1985, the urethane foam expert at the National Bureau of Standards reported on his tests. He had experimented with various formulations of urethane foam and concluded that the fire retardants and other variations of formulation had little influence on the smoldering of the foam. Smoldering foam poses a major toxic risk. It is obvious that the Urethane Safety Group deliberately fostered confusion over toxicity and flammability, the two different but related risks of the foam. This is false and misleading advertising. When asked about fire deaths from smoke inhalation, the group stated that the foam is fire retarded and won't burn until a high temperature is reached. With this response, the group is essentially giving out false information, even if the statement is true. The question asked was about toxicity, which occurs when the foam *smolders;* the answer that was given mentioned only *flammability*. The fire retardants actually add to the toxicity of the decomposition/combustion products. The implication of the group's statement is that fire retardants will prevent the fire deaths connected with urethane foam. This is misleading and deadly.

The group also pulled together a tremendous lobbying effort to prevent government agencies from banning the use of the foam in mattresses and stuffed furniture. Much of this lobbying effort made use of the same slick misleading statements that were also put into public relations. Typically, only nontechnical administrators would be lobbied, since most of them didn't know enough to question the statements.

## THE VINYL GROUP

At a meeting held in Cleveland in 1976, the Urethane Safety Group was consciously chosen as the model for the Vinyl Group. The Vinyl Group met to decide what to do about the rising criticism of PVC in fire deaths and injuries. They decided to "rehabilitate" the product, primarily by using public relations efforts similar to those used for urethane foam. Many of the arguments that had been used for urethane were also used for PVC: confusion over toxicity and flammability, statements that all organic materials are toxic when they burn, and the assertion that carbon monoxide is the lethal chemical in all smoke, so PVC is no worse than any combustible material.

The Vinyl Group eventually became the Vinyl Institute, which has a "truth squad." Whenever a fire involves PVC as a contributing factor to deaths and/or injuries, the "truth squad" denies the involvement and points to another contributing factor as the cause. We all know that fires are complex events, with many factors contributing to the outcomes. Any person or group discussing the cause of deaths and injuries at any one fire should be immediately suspected of great bias. The Vinyl Institute churns out misinformation and misleading statements at a rapid rate.

## OTHER CASES OF MARKETING AROUND HAZARD

Other groups that have used these techniques as a model for their own products include the Plastic Pipe Institute and manufacturers of PVC electrical conduit. The techniques spilled over into the realm of dangerous nonplastic products. For example, in about 1982, the public relations firm of the Society of Plastics Industries submitted a proposal to the Kerosun Corporation, manufacturers of kerosene space heaters. The firm explicitly recommended that Kerosun use the plastic pipe campaign as a model. The proposal called for a "truth squad" to run around and talk about non-kerosene heater causes in kerosene heater-involved fatal fires. It also suggested putting the emphasis on economics, and sweeping health and safety issues under the rug. It mapped out a massive lobbying effort to soften local and model codes to allow kerosene heaters in homes. The proposal was, essentially, a perfect example of marketing around hazard.

Other products being marketed in this way include polyethylene shipping containers for combustible and hazardous materials, plastic automobile parts, fiberglass underground fuel tanks, vinyl siding on homes, and weight-bearing composites (these are layers of different kinds of plastics being proposed for small bridges, piers, and houses). These methods of marketing and of lobbying will result in many, many deaths.

## THE EFFECT OF THE SETTLEMENT ON PLASTICS RESEARCH

The FTC settlement with the urethane foam group provided the motivation for the plastics industry to flood with its influence the flammability and combustion toxicity research structure. Academic programs became huge, with centers of fire research established with plastics industry money. Internal "research" was pursued in plastics corporation laboratories. As has been stated before, experimental design for much of this research biased the results. This bias can be perceived only by a trained scientist familiar with the issues in question and with the experimental apparatus used.

The plastics industry also funded much of the research carried out by the National Bureau of Standards' Fire Research Center. The industry established fellowships within the Bureau and installed its designated technical personnel there. At first, rather fundamental problems in fire behavior were explored—fairly noncontroversial topics. Later, the toxicity issue was addressed, and what emerged was a toxicity test that cannot discriminate between materials. (The test procedures render the toxicities of most materials essentially the same. Wood tests out as toxic as rigid urethane foam.) The Bureau now tests and tests all kinds of materials and regularly ends its report with a phrase such as, "This material is no different in toxicity from anything else." With such a test, of course, the results would show no difference. Very few materials show unusual toxicity when tested by the NBS combustion toxicity test.

The Consumer Product Safety Commission relies on the NBS to test for fire hazards of products, including toxicity. It also has begun relying more and more on volunteer programs established by the regulated industries. By voluntarily putting certain warning labels on products, manufacturers escape stiffer regulations. The labeling of stuffed furniture with warnings of flammability and possible toxicity in the event of fire is an example of a volunteer program whereby the industry fended off good regulation of a hazardous product.

The EPA gained notoriety in the days of Rita Lavelle, the EPA official who was found to have given special favors to industry in 1980–1982. She was imprisoned for this. During the time she was in office, there were cozy meetings between members of the toxic substance bureau and industry representatives. In regulating plastics as toxic substances, the EPA bent over backwards to have polymers declared nontoxic. In 1982, public hearings were held to delete major classes of polymers from the requirements of the Toxic Substance Control Act. The Federal Register notice of the proposed rule changes included masses of industry data and very little about the negative environmental and health impacts of these polymers. The notice also asked for data to widen the easing to plastics based on halogens and nitrogen. The notice denied that plastics were toxic, ignored the fire problem, ignored air pollution problems from disposal of plastics, and ignored the problems caused to wildlife by plastics disposed by careless people. This notice, and the conduct of the hearings, formed a clear example of deliberate and studied ignorance. This is inexcusable, especially when the public welfare is involved.

## OTHER RESULTS OF LOBBYING

The plastics industry successfully lobbied to the detriment of public welfare over and over again. Results of this lobbying included deceptive wording of solid waste laws and the handing over to the National Sanitation Foundation the work of generating regulations for drinking water laws involving chemicals from pipes and from treatment plants. The National Sanitation Foundation purports to be an impartial environmentally concerned foundation. However, it is almost entirely funded by the Plastic Pipe Institute of the SPI and by plastic pipe corporations.

Other federal agencies that give every sign of having been lobbied by the plastics industry and unduly influenced by such lobbying include the Department of Transportation, in its regulations on shipping containers; the Fire Data Center of the Federal Emergency Management Agency (FEMA); and the National Research Council (NRC) in its report on Fire and Smoke Toxicity. The Council's mishandling of fire fatality data (use of gross fatalities instead of fatalities per fire) is the same as that of the plastics industry's public relations campaign. Some of the steps in logic also echo and have the same fallacies as the house of logic built by the SPI. The befuddling of such a prestigious scientific body as the National Research Council is particularly sad.

To gain an idea of the full range of governmental and quasi-governmental agencies and other regulatory and research institutions that were either deflected from their objectivity by plastics industry funding or whose results were misused by the industry, see the November 1976 issue of *Fire Journal,* which contains an article by SPI executive John Blair. This article was written to extol the breadth of the industry's research efforts. But now that we have seen the results of these efforts, all the article does is expose the size of the octopus.

## INFILTRATION IN CODE MAKING

The National Fire Protection Association began as a combined effort of the fire service and concerned citizens. It was formed to combat the truly horrendous loss of life and property caused by fire in this country. Even today the United States ranks first out of all the developed countries for both fire fatality rate and fire damage.

One of the major avenues of combating these losses has been the set of codes implemented by the NFPA. There is a committee responsible for each of these codes. These committees are responsible for considering changes in the existing code and proposals for new code sections. Codes are adopted by consensus. Committee members are supposed to avoid conflict of interest by not representing industries directly affected by the code being considered.

In discovery papers received from the aluminum wire manufacturers and the manufacturers of PVC wire insulation, the counsel for plaintiffs in the Beverly Hill Supper Club fire (in Covington, Kentucky) found that these committees work in less-than-ideal ways. Members of the committee on electric wire insulation included representatives of the aluminum wire industry, and members of the committee on electrical conductors included representatives of the PVC wire insulation and larger PVC manufacturing industries. The aluminum wire and PVC wire insulation industries are sister industries in the manufacturing of aluminum electrical wiring. The committee members would be overtly lobbied by the regulated industry and covertly lobbied by the representatives of the sister industry who were members of the code panel. The aluminum wire industry representatives and the PVC industry representatives had made a deal. This is the way code making by consensus was subverted and became a means of endangering the public.

In reacting to rising demand from firefighters and the public that the issue of combustion toxicity be addressed, the NFPA established a toxicity committee in 1982. The committee was heavily weighted with scientists either directly or indirectly employed by manufacturers, or large users of plastics, such as the airline industry. The committee concluded that not enough was known about the combustion toxicity of the different materials to take any action. It recommended further study. Indeed, no action has been taken by the NFPA on the issue since then. What a far cry from the President's Commission on Fire in America, which a decade before had remarked upon the unusual toxic hazards posed by plastics!

The plastics industry infiltrated the NFPA in other ways. Joint sponsoring of fire toxicological research was one way. NFPA lent its credibility, but SPI influenced the design of the studies and the wording of the resulting reports and scientific papers. While the plastics industry infiltrated NFPA's toxicology committee in two ways, the major consultant of the committee was also under SPI funding for other work. Also, employees of several major users, like the air travel industry, sat on the committee.

Another committee slightly infiltrated was the fire service committee, which was made up of firefighters and fire officers. At least one fire officer serving on that committee took SPI money to watch the activities of fire service personnel and citizens who wanted to regulate plastics for fire safety reasons. He also lent his credibility as a fire officer to public statements in support of plastics. When this officer's activity and source of funds was discovered by his union local, he was severely chastised by the executive board of that local. Fire officers know about the properties of plastics in fires and campaign against the unnecessary and unsafe use. The subversion, or at least attempted subversion, of the NFPA fire service committee by the SPI, as well as these other acts of infiltration, demonstrates the urgent need to reform fire and building code-making practices in this country.

Other model code-making bodies have been similarly infiltrated by the plastics manufacturing and using industries. An Underwriters Laboratories listing, which is supposed to indicate fire safety in all respects, has even been allowed for products that the laboratory has never tested toxicologically. And the flammability tests for these products were designed by the product manufacturers. UL has been named a defendant in several cases of fires involving PVC electric wire insulation. The UL label on a product does not necessarily mean that the product is safe for reasonable use.

These model codes are adopted by local governments, which have their own code-making and code-adopting bodies. When Hill and Knowlton proposed that Kerosun try the same successful local code campaign that they employed for plastic plumbing pipe, they suggested the use of slickness, economic arguments that ignore safety concerns, a "truth squad" to confuse the issue in the aftermath of a fire involving the product, and media hype.

**What Happened in New Mexico**

The Carlon Corporation, which makes PVC electrical conduit, tried to get the electrical code changed in New Mexico and worked with (and on) a not-too-bright state official in charge of the electrical code. Because of the clumsiness of both Carlon and this official, the anatomy of the lobbying and code-change became too exposed. In fact, the governmental pawn actually made the mistake of naming Carlon in the proposed code change itself! That mistake resulted in the rejection of the change by the electrical code council of the state.

The official tried a second time. This time his mistake was to give a newspaper interview that clearly showed that he was not objective and was still dealing with Carlon and possibly other corporations. This interview resulted in his disqualification as the presiding officer for the hearing on the proposed code change.

These hearings in New Mexico featured lengthy testimony by a large number of national experts. Some of these experts were financed by the plastics industry, some by the steel industry, and some by litigants in the large lawsuits. Two experts giving testimony at these hearings were Ted Radford of University of Pittsburgh and Irving Einhorn, formerly of University of Utah but now an independent consultant. These experts were paid by the plastics industry to testify. They had infiltrated the plaintiffs' experts team of the MGM Grand Hotel fire litigation. When a tape of their testimony was sent to the Plaintiffs' Legal Committee, their bias and their conflict-of-interest were so outrageous that many of the lawyers wanted to sue them also.

The electrical code council of New Mexico did not buy the slick show put on by the plastics industry witnesses. The code council saw through the obfuscation, half-truths, facts taken out of context, and appeals to economics. The change in the code itself was narrowed to allow PVC electrical conduit in an extremely narrow range of buildings; indeed, the flexible PVC conduit was limited to a building with all the attributes of a blockhouse.

In New Mexico, clumsiness rescued public safety and health. No accusation is made here that this official took money or received luxury goods, sex, or holiday trips. Corruption may include these kind of goodies, but more often it involves flattery, ticklings of the ego, and giving the impression of competence to incompetent decision-makers.

## What Happened in New York

A second example of greater-than-usual publicized lobbying by the plastics industry occurred when New York State tried to amend its Building and Fire Protection Code to include a combustion toxicity data bank. I was directly involved in this process and saw this lobbying first hand. The first round of hearings occurred before the Senate Committee on Insurance and concerned the funding of a feasibility study on decreasing combustion toxicity. Many of the same experts who testified for the plastics industry in New Mexico testified in Albany as well.

During these hearings, a new element was introduced. Professor Emmons of Harvard wanted all attempts to regulate combustion toxicity halted until his computer model was perfected. This hearing took place in 1982. By 1985, Professor Emmons was retired and the model was still unperfected. The cry for an integrated hazard assessment model was taken up by both the National Bureau of Standards and the National Research Council's Subcommittee on Fire and Smoke Toxicity. (An integrated hazard assessment model is a model that includes all aspects of fire hazard for a particular situation. It would include building design, furnishings, and occupancy. The aspects of a fire hazard include fire spread, heat, density and toxicity of smoke, and so forth.)

Again, the decision-makers did not believe the slickness and voted for the feasibility study. This study, completed in 1983, recommended establishment of a combustion toxicity data bank based on Yves Alarie's test protocol, usually called the University of Pittsburgh test. (See Chapter 2 for a description of this test.)

The study report and recommendations were presented by the Secretary of State to the Code Council in 1984. This presentation followed a year of intense lobbying and political effort by the plastics industry on the one hand, and the environmental groups and fire safety groups on the other. The industry used all its usual weapons—invitations for "joint studies" financed by the industry, threats of lawsuits, attempts to change the operations of the code council by legislation with the help of a few bought senators, and attempts to get the Secretary of State and key assistants fired.

The efforts of the industry delayed the final approval of the amendment by three or four years and weakened it severely. The amendment now covers only building systems and interior finishings, not furnishings or exterior finishings. This change occurred as a result of the regulatory impact assessment, which was performed to put off one of the lawsuit threats. A **regulatory impact assessment** or analysis is an estimate of the impact of a proposed regulation on the economy and on the problem for which the regulation is written. It is like a cost/benefit analysis.

The industry looked hard for a weak point on the code council and found one in the Department of Housing and Community Renewal, the lead agency of the code council. The staff of the council was drawn from that Department and voiced all of the industry's concerns, but virtually ignored all the concerns of the fire safety groups. Although there is no hard evidence, as we had for the fire officer, that money changed hands, these staff members used phrases straight out of the SPI public relations handouts and instruction sheets. These staff members formed the industry's conduit into the code council, for whatever reason. They were responsible for a large proportion of the delays in implementing the data bank, including a last-minute attempt to change the basis of the data bank to the NBS toxicity test, instead of the Alarie (University of Pittsburgh) test. They wanted to "stiff" the people of New York with an inferior test method just to please their industrial masters.

## OTHER HAZARDS OF PLASTICS

We have been describing the fire safety aspects of regulating (or not regulating) plastics. Plastics cause other environmental and health problems besides those associated with fires.

- The byproducts of the manufacturing of plastics are toxic.
- The manufacturing involves air and water pollution.
- Use of the products may result in indoor air pollution from slow, inexorable decomposition.
- Wildlife is dying from eating plastic or getting entangled in it.
- Chemicals leach out of plastics in landfills and into surface and ground water.
- Chemicals also leach out of plastic plumbing pipe and into drinking water.

At every life stage, plastics impact the environment or health.

There are a large number of federal agencies that are supposed to regulate these impacts. The EPA has charge of drinking water standards, toxic wastes, air and water pollution, and wildlife impacts. The Department of Transportation (DOT) is in charge of the use of plastic as shipping container materials and as construction materials for vehicles and airplanes. Housing and Urban Development (HUD) has charge of housing codes for federally subsidized housing. The Consumer Product Safety Commission (CPSC) issues regulations with regard to health and safety hazards from consumer goods, such as the use of phthalate in plastic baby toys. (Phthalate is a carcinogen and cardiotoxin.) With all of these agencies supporting their bureaucracies, we might expect some protection. Great expectations! Let us consider the EPA as an example.

**What the EPA Has (or Has Not) Done**

In recent years, the EPA has performed a number of "kindnesses" for the synthetic polymer industry. The EPA held hearings in Washington in 1982. The purpose of these hearings was to remove those polymers that do not contain either a halogen or a nitrogen from under the reporting requirements of the Toxic Substances Control Act. Thus, these polymers would essentially be labeled nontoxic. The EPA used industry-supplied data for the foundation of this rule. Data in other bureaus in the EPA contradicted many of the assertions made of the preamble and conclusions of this proposed rule, such as data on indoor air pollution, on the leaching of chemicals into water and their effects on aquatic life, and on pathways of exposure to polymers. Furthermore, the EPA concocted some incredible scenarios about exposure to polymers. For example, they estimated exposures to paint at ridiculously low frequencies and durations. Those of us who do our own painting have considerably greater exposures than the estimates. Obviously, the Bureau of Toxic Substances at the EPA was not purged when Rita Lavelle went.

Very recently, the EPA placed the National Sanitation Foundation under contract to produce standards for the use of polymers in water treatment plants and in plumbing pipe. This foundation is largely funded by plastic pipe manufacturers and has a long history of fronting for the industry. The EPA essentially put the fox in charge of the chicken coop.

The EPA is also guilty of the following:

- Reliance on DOT regulations for containers of hazardous materials in setting standards for containers for hazardous waste.

- Attempts to loosen vinyl chloride monomer limits for air releases from PVC factories.
- Complete failure to look at wildlife involvement with plastics, until very recently.
- Complete failure to act on the problems caused by plastics in municipal solid waste.*
- General failure to conduct objective, noninfluenced reviews of environmental impacts of plastics.
- Failure to investigate the role of plastics in indoor air pollution.

Manufacture, use, and disposal of plastics also contributes to the two major hazards of the global atmosphere that concern environmental scientists—the eroding of the stratospheric ozone layer, which results in more of the sun's harmful rays reaching the earth, and the "greenhouse effect." The greenhouse effect is a global warming that is the result of the use of fossil fuels, forest fires, methane from animal waste, and other processes. The elimination of the earth's rain forests also contributes to the greenhouse effect, because these trees are necessary to clean the atmosphere. During the manufacture of plastics, molecules containing the poisonous halogens fluorine, chlorine, and bromine are released into the atmosphere, contributing to the erosion of the atmosphere and the poisoning of the air. This release occurs more slowly during normal use and decomposition-aging of the plastic. If the plastic is incinerated, the release is again rapid.

In 1988 an international treaty was signed in Montreal to concentrate on the elimination of the use of chemicals that

---

*Plastics endure for decades in landfills because they don't biodegrade (decay because microorganisms break them down). But they become unusable fairly rapidly because of chemical decomposition (the loss of certain parts of the molecule). Because of chemical decomposition, plastics are quickly unusable, but discarded plastic is non-biodegradable! This obviously contributes to the trash disposal crisis.

harm the atmosphere. The eighty countries involved were brought together by the United Nations Environmental Program. One of the culprits singled out by those who signed the treaty was the freon blowing agent that is used in the manufacture of many urethane and polystyrene foams. Depletion of the ozone layer would slow down if this substance were eliminated.

The manufacture, use, and disposal of plastics also generates the small organic molecules, such as methane and much carbon dioxide, both of which contribute to the greenhouse effect. In general, the use of fossil fuels such as oil and coal results in worsening the greenhouse effect, but few people realize that the manufacture, use, and disposal of plastics, solvents, pesticides, and glues also employs the use of fossil fuels.

The EPA has not reviewed these atmospheric impacts of plastics, except for freon blowing agents. The EPA has been very kind to the plastics industry.

## HOW INFILTRATION IS ACHIEVED

Both the individual corporations of the plastics industry and the umbrella trade association, the Society of Plastic Industries, have perfected lobbying and infiltration techniques often used by other industries. Little meetings and luncheons are held in private with governmental and code-making body officials. These meetings foster a sense of equality, identity, and power-sharing.

If the nice-guy techniques don't work, threats are used. The carrot-and-stick method also has been used in the form of threats and promises of goodies such as political support for a higher position or the promise of a job in the industry.

Regulators need a technical veneer to parade before the public so industry provides masses of "designed" data. These data are the result of carefully engineered experiments that appear to show that plastics pose no health hazards. But these experiments were actually carried out under outlandish/unrealistic lab conditions and often the data are inappropriately analyzed and interpreted. Another tactic that is used by technical support to weaken regulation is the pointing out of the prevalence of these materials in the present environment and listing the prestigious governmental and quasi-governmental agencies that have already approved the use of the material.

Some regulators actually begin to feel more loyalty to the material and the industry than to the public and to the governmental charter. A good industrial lobbyist establishes both personal relations with the regulators and professional teamwork. The lobbyist emphasizes a sense of belonging to an elite group. Social network, group security, and ego stroking can corrupt as surely as money and other concrete offerings can and the industry has also been known to offer all of these in large quantities. The nonconcrete may be more damaging to public safety in the long run, because corruption in the broadly understood sense cannot be proved. We shall enlarge upon the social and psychological aspects of corruption in the next chapter.

## COURT SETTLEMENTS

We have discussed how regulators and legislators try to hide the threats that plastics pose to the environment and to health, and the effect that this type of coverup has on public safety. But another branch of government is also important in its effects on regulation, namely the courts. A large number of plastics-related cases are in the courts at any one time. The majority of these cases involve deaths and injuries from fires, but some touch on other aspects, such as the dumping of the toxic by-products left over from both the manufacture and the physical failure of the plastic. Most of these cases are settled out of court, before the judge or jury can hear the evidence in open court and make a decision. In this way, the corporations can not only avoid judgment but can keep the evidence from becoming part of the public record.

In many cases, the corporations require a special condition for settlement—sealing of the record. By sealing the record, facts about the circumstances of deaths and injuries are not accessible and are therefore kept out of the public eye.

After a case is settled and its records are sealed, the plastics industry releases statements that no deaths or injuries have ever been connected with the product. In most cases, these statements are lies. What is frustrating for those who know the truth is that, because of the sealed records, the facts cannot be exposed. The federal regulatory agencies and the code-making bodies do not look into the sealed records—indeed, they *can't* look into the sealed records. The decision-makers cannot have access to the data and information that they need in order to render a decision that is technically sound. As long as the industry can pay the settlements, the hazards to the environment and to public health and safety will continue.

## THE RESULTS OF INFILTRATION

The infiltration of these bodies and government agencies by plastics manufacturers and trade associations is possibly illegal. The results of these actions are certainly dangerous. They include the weakening of existing protective regulations, codes, and standards, and the failure to promulgate newly needed regulations, codes, and standards. As a result, the utilization of existing plastics widens, and new plastics are introduced on the market. Health, safety, and the environment are then threatened by every stage in the life cycle of these products: manufacturing and processing, transportation, storage, use, and disposal. The impact of the plastics include the following.

- Contamination of environmental media (air, water, and soil).
- Involvement with wildlife (starvation by animals who mistake plastic for food, and the entanglement of animals in the products).

- Structural failures by physical degradation.
- Deterioration of consumer and housing economics.
- Leaching of chemicals from plastics in landfill.
- Toxic pollution in landfilled ash and in flyash from incineration of plastics.
- Indoor air pollution from decomposing plastics in "tight" buildings.
- Problems associated with plastics in all kinds of fires (factory, warehouse, residential, hotel/restaurant, and office building, as well as rubbish).

Plastics impact the environment and society in so many serious ways. Eventually, society pays for "cheap" materials in one way or another.

## SOLVING THESE PROBLEMS

The only effective paths to solving these problems are integrated approaches. Integrated means that the problems caused by the products are *not* addressed one-by-one; instead, the total impacts are reviewed together, and solutions are developed that minimize as many of the impacts as possible. One integrated approach is the minimization of the manufacture and use of these plastics. The other is by integrated regulation of all of the life stages of plastics, from manufacture to disposal, in a coordinated and nonfragmented way. Both of these approaches imply that neither infiltration nor ethically dubious cooperation would be allowed to interfere with the effectiveness of these programs. Such an implication means that sunshine laws would be in operation and the "little private meetings" would be forbidden. It also means that dedicated environmental and public health groups would be expected to assign individuals or committees to watchdog these programs.

*Chapter 10*

# Dragonslayers Sharpen Swords and Pencils

While those in charge of manufacturing plastics and regulating their use work together to soften the regulations, more people die or are injured. Since the plastics industry possesses great wealth and power, what can be done to curb the inappropriate and dangerous uses of plastics? If the plastics industry hires armies of glib public relations wizards (and it does), how can the public be educated honestly and impartially about the dangers and the environmental impacts of plastics? Since the plastics industry sends swarms of lobbyists to elected officials, regulators, and code-making bodies, how can the laws, regulations, and codes be changed to reflect the dangers that plastics have for our environment and our health?

The plastics industry is indeed wealthy and powerful, but it is also like the emperor in the tale *The Emperor's New Clothes*. Every so often, an ordinary citizen carefully watches and listens to a public relations wizard and sees through the clever phrasing. Once in a while, a lobbyist from a rival industry catches a regulator who has been involved in dishonest dealing with the plastics industry, and the exposed culprit then hangs out to dry for the edification of the public. Even with all of its wealth and power, the plastics industry faces one limitation: it can't change reality.

Suppose a citizen decided that the dangers from inappropriate uses of plastics greatly exceeded the benefits and that Something Had to be Done! The political and social archives of civilization bulge with accounts of individuals and groups who started, with good intention, To Do Something. Presently, widespread woes that need solutions include homelessness, substandard housing for poor and working-class people, serious atmospheric contamination, and weapons capable of massive destruction that are stockpiled in many countries' arsenals. Well-intentioned groups have mounted their best efforts to combat these woes, with ambiguous results. The citizen who decides To Do Something must analyze and define the problem and then not Do Something, but Do *The* Thing—that is, identify and attack the vulnerabilities of the problem as they emerge from the analysis, just as St. George attacked the dragon as it emerged from its lair.

The previous nine chapters presented the problem of plastics in fires in great detail. A summary of some of the aspects of those details, presented analytically, would help the concerned citizen structure an effective dragonslaying program.

## THE EFFECTIVE PROGRAM

An effective program builds on the pillars listed below.

- The characteristics of the materials to be regulated.
- The circumstances in which these materials have proven themselves dangerous.
- The dynamics of the political and social system.
- The attributes of individuals and groups which empower them (such as imagination, commitment, persistence, capacity for indignation, discernment between truth and obfuscation, and fellowship, etc.).

The first two pillars dictate the goals and objectives of the program, the third pillar determines the particular vehicles for obtaining the goals and objectives, and the fourth pillar achieves the authority for imposing the vehicles.

A short digression about this hypothetical citizen. The person who ultimately initiates and propels the action in an important issue is largely unpredictable with respect to gender, ethnicity, age, level of education, or level of income. However, this person does have certain predictable virtues: anger, rather than despair, over the endangering of human life and coverups that continue the danger; a desire to communicate the anger rather than bottle it up; a high natural intelligence and human sophistication; curiosity; the willingness to learn and the ability to hear; and deep empathy. This is the person who can build on the four pillars. Now let us review three of the four pillars.

## The Inherently Dangerous Characteristics of Plastics

The synthesis of plastics requires large quantities of energy in the form of heat. This becomes incorporated into the chemical bonds of the molecules and is transformed into chemical energy. This high energy content runs counter to the thermodynamics laws, which favor low energy, and renders most of the commonly used plastics inherently unstable. The energy level of the molecule is so high that the molecule tends to fall apart. This high amount of energy means that even during manufacture, low level decomposition occurs, resulting in defect points in the resin. Nearly all plastics on the market have defect points in them from manufacturing and processing.

Plastics are not solids like glass or metal. They are viscoelastic fluids. Their physical and chemical properties are different from true solids, but they are marketed as if they behave just like the solids which they replace. This kind of marketing has serious consequences.

Plastics age in ways that differ from the aging of solids and of low-energy natural polymers, like wood, cotton, silk, etc. One form of aging is called "creep": the plastic slowly flows. Figure 1.1 illustrates this. Eventually, "creep" affects the strength and other physical properties of the plastic and its chemical stability. Other forms of aging include distortion, expansion, oxidative embrittlement, stress cracking, and chemical stress. Environmental factors determine the rapidity and form of the aging of any particular piece of plastic, although each particular plastic is more prone to a certain form of aging than another.

Because of these high-energy chemical bonds and defect points, even mild environmental factors induce chemical degradation in most common plastics. Under normal use, most plastics slowly emit pieces of the molecule that are not part of the main carbon skeleton. In many plastics these pieces are irritants. In addition, most plastics emit their additives, the chemicals (such as fire retardants, plasticizers, and colorants) added to the plastic that do not become part of the polymer molecule itself but modify its properties. One of the most common organic air pollutants today is the plasticizer phthalate, because it is used in a wide variety of plastics and is easily emitted. When the sun beats down on your car and you open the door, you usually smell a "plastic" odor, which is a product of the slow degradation and the additives. Most of us are getting doses of irritants, hydrocarbons, and additives all the time from "innocent" actions like parking cars in sunny places.

In a situation of intense heat, the instability of the chemical bonds can be dangerous, because the breaking of the bonds greatly accelerates and leads to greater emission of the additives and pieces of polymer molecule. Each plastic has a characteristic temperature (temperature of quantitative decomposition) at which large and predictable quantities of these chemicals are emitted. For many of the commonly used plastics, the emission rates at the temperature of quantitative decomposition are high enough to cause toxic concentrations of these chemicals. For PVC, clouds of hydrogen chloride are emitted; for urethane foam, the emission is a mix of narcotizing and irritating hydrocarbons; and for nylon, hydrogen cyanide and nitrogen dioxide are major decomposition products. The temperature of quantitative decomposition for plastics is usually much lower than the ignition temperature. People are exposed to these concentrated emissions without even knowing that fire is in its early stages—no flame is present.

It may take a long time for a flame to appear, and the exposure may continue for some time. When the fire does ignite, the combustible gases emitted during decomposition flare rapidly, and the fire spreads quickly. Because of their greater carbon density and fire retardancy, plastics emit more carbon monoxide during fires than wood or other natural polymers.

The scenario accepted by most fire scientists is that during quantitative decomposition people are exposed to chemicals that injure them and render them unable to escape. They then inhale the high concentrations of carbon monoxide during the combustion stage and die as a result of the combined dose.

## How These Materials Have Proved Dangerous

These materials are dangerous enough alone, but when combined with other materials, or stored in places without proper ventilation and without consideration of fire safety, they become even more deadly. The way a building is constructed will have some affect on the outcome of a fire.

### *How Fires Behave in Buildings*

The built environment contains numerous sources of energy that can cause smoldering and/or burning of combustible materials. Before smoldering or burning occurs, an energy source will release excessive heat into an enclosed space for a period of time. Heat will accumulate on the ceiling and on surfaces near the energy source. When the heat flux becomes great enough, a material will reach its temperature of quantitative decomposition and emit gases. If heat further accumulates, the gases or the material itself may ignite. In most fires the primary fuel itself does not burn, but its decomposition gases do.

Once the object reaches either open flaming or propagation smoldering, it begins to *release* heat, as opposed to the earlier heat *absorbing* stages. It heats other objects by irradiation and convection, and spreads the fire.

Fires establish an airflow in which heated air and products of combustion rise and spread out below the ceiling while cooler air flows toward the fire along the floor, as seen in Figure 2.1. Two atmospheric layers are established. As the fire grows and consumes more fuel, the upper layer expands and the lower one shrinks.

As long as the plume from the fire is hotter than the air around it, it will rise and spread along the ceiling. Openings between floors spread the gases and smoke from one floor to another. These openings can be elevator shafts, electrical outlets, open windows, plumbing openings, and even the air handling system. Gases and smoke will spread laterally and vertically under ceilings. If no openings are available laterally or upward, the products of decomposition and combustion will spread downward as in the fire at the MGM Grand Hotel.

Very typically, the fatality victims and injured survivors of a fire are found in a location that is distant from the fire's origin. Often they are not even near any of the flames. The design of a building can facilitate the spread of lethal and injurious concentrations of smoke and gases far beyond the area directly affected by flames. The operating mode of the air handling system during the fire, as well as the design of the building, will determine the pattern of smoke spread.

In summary, because of the enclosed nature of buildings, the following is often true.

- Most fires in buildings subject fuels to rising temperatures.

- Nearly all fires in buildings involve flows of air and of products of decomposition and combustion in two layers.
- Materials subjected to rising temperatures begin emitting products of decomposition before they ignite; indeed, decomposition gases are usually the actual fuel that burns.
- The pattern of smoke spread within the building determines the injuries and deaths and may reach far beyond the area of flame spread.

### *Specific Fires*

The detailed reviews of the fires in Chapters 3–8 provide case studies for the general principles governing the characteristics of plastics and of fires in buildings. The reviews reveal the types and quantities of plastics that are commonly involved in fires: PVC, urethane foam, nylon carpeting, styrene-containing plastics (ABS, polystyrene), acrylics, and polyolefins (polyethylene and polypropylene). In large buildings, tons of these materials may be present. Hundreds of feet of these materials may stretch in an uninterrupted length as carpeting, wallcovering, electric wire insulation, or plumbing pipe.

The chapters describing these fires show that ignition of these materials can come from rather ordinary sources, such as an overloaded or slightly faulty electrical system, a carelessly thrown cigarette, or human carelessness with flammable fluids. There is no dearth of ignition sources, and it must be assumed that a fire can happen in any building. The fire can ignite at any time of the day almost anywhere.

We also saw that the design of the structure of the building interacts with the types, quantities, and placements of the plastics to determine the pattern of smoke and fire spread. Tall buildings, like the MGM Grand Hotel and the New York Telephone Exchange, usually act as chimneys, drawing the smoke upward to accumulate under the roof. Corridors and subway tunnels may behave as sideways chimneys because the air and smoke must essentially flow in only two dimensions, not three. The fires in the Fort Worth Ramada Inn, Westchester Stouffer's Inn, and the PATH tunnel under the Hudson illustrate this "sideways chimney" effect.

The analysis of the MGM fire shows that continued operation of the air handling system can help spread the smoke to distant parts of the building. In fact, structures that run from bottom to top and have no smokestops will carry the smoke very rapidly to the top of the building through elevator shafts, defective firestairs, seismic joints, and plumbing and electrical systems.

Chapters 3–8 show that once people are exposed to the smoke for more than just two to three minutes they have great difficulty escaping. Some of these fires occurred in large structures such as skyscrapers and tunnels, from which escape is inherently difficult. But the Ramada Inn was only two stories high; the Stouffer's Inn conference center, three stories high; and the Younkers Brothers Department Store, two stories high with many exits. In these three fires, those who survived got out within two to five minutes of smoke exposure. As revealed by the Younkers Brothers Department Store fire, incapacitation can be nearly instantaneous if the smoke is highly concentrated.

We won't review the characteristics of the deaths and injuries here except to note that they were much different from those seen in fires fueled by traditional materials. The presence of the corrosive irritants caused much tissue damage along the respiratory tract and, in injured survivors, problems of the circulatory system. Cyanide was a major toxin in the smoke. The hydrocarbons caused neurological, reproductive, and skin problems in the injured survivors. Because of the presence of plastics, the probability of death or injury from smoke inhalation has changed, as has the form of death and injury from smoke. A high proportion of the injuries are permanent, and the lives of the injured survivors are changed dramatically.

## The Dynamics of the Political and Social System

Chapter 9 discusses how the use of plastics came to be so prevalent in buildings. In brief, every governmental and code-making entity entrusted with the task of public health and safety in buildings has failed the public. At the Federal level, agencies with jurisdiction include the Consumer Product Safety Commission, the Fire Research Center of the National Bureau of Standards, the Federal Trade Commission, the Environmental Protection Agency, the Occupational Safety and Health Administration within the Department of Labor, and the National Institute of Occupational Safety and Health within the Public Health Service (Health and Human Services). Also, the environmental impacts of manufacturing, using, and disposing of plastics are, at best, poorly addressed by the EPA, and at worst ignored.

Local governments (state, county, and municipal levels) are heavily lobbied by the plastics industry to allow more and more substitution of noncombustible materials and natural polymers by plastics. Thus, large areas of the country, especially in the South and West, have many buildings with plastic plumbing and Romex electric cable sheathing. The public relations firm of Hill and Knowlton, often used by the plastics industry for this local lobbying, set down in writing the strategy for this effort: gloss over the safety questions, but talk up economics.

The National Fire Protection Association develops most of the fire-related model codes in this country, although other code-making bodies also establish relevant model codes. The system of committees and voting used by the NFPA leaves the code-making process open to abuse, and abuse does occur. In the past, deals were made between industries serving on various committees, and the voting process was tainted by industry lobbying. At least once the fire service committee was revealed to be infiltrated by a fire officer who was accepting money from the Society of Plastics Industry. When the plastics industry misuses its power and money this way, the public is not served.

The NFPA staff has investigated multifatal fires, but its investigations have been technically inadequate, and they rarely discuss the role of the particular mix of fuels in the toxicity of the smoke. The staff appears fixated on attributing fire outcomes to NFPA code violations. In the case of the Younkers Brothers Department Store fire, there were no serious violations, and the NFPA investigator reported that he could not draw any conclusions about the cause of the fire.

By the late 1970s, the NFPA was jointly sponsoring studies with the Society of Plastics Industries. Not surprisingly, most of these studies yielded equivocal results about the role of plastics in smoke toxicity. The low technical competence and the outdated understanding and attitudes of the NFPA staff toward the various influences on fire outcomes appears to have made them unequal to the task of working with the SPI in designing, managing, and interpreting these studies.

The bottom line is that the activities of the NFPA in code-making, fire investigation, and fire research are not serving the needs of the citizen or of the firefighter in the area of smoke toxicity and fire behavior of plastics.

The fourth pillar refers to the attributes of those who must fight the dragon. This is described in the next few sections. Also discussed are the methods that can be used in the battle.

## HOW CAN WE DEFLATE THE DRAGON?

The solution to these problems is more difficult now than it used to be, for several related reasons. The influence of money has increased greatly, even with local politicians such as state and municipal legislators. Thus, the number of potential votes needed to draw the legislators' attention to real problems increases in direct proportion to the volume with which money talks to these legislators. The ordinary old-fashioned ways of communicating with legislators (letters and phone calls) frequently have little effect, but must be used anyway.

Both legislators and regulators also seem to have progressively contracted a disease of ego, cosmetic actions, and big media campaigns, with none of the follow-up necessary for problem-solving. Getting appointments becomes more difficult under these circumstances. Getting commitments for laws and regulations is even more difficult. The legislator or regulator may even have a big press conference and discuss the risk to life and health that plastics pose. For some of these officials, such a press conference is a signal to the public that nothing real will be accomplished.

Another necessary but insufficient route is public education through the mass media. There are many people who believe that if the newspapers would print a story or if the television stations would run an exposé, then the citizenry would be aroused and everything would be solved. Unfortunately, all that can be expected is that the public will be somewhat educated about plastics through the mass media. At this time in mass media history, the public has been saturated with the Geraldo Rivera-style of journalism: too much sensationalism, too much blood, too much scandal, and too much kinkiness.

Even when the public *does* react to such grave and real problems as plastics in fires, organizations must exist to channel the reaction. Without problem-solving programs in which an educated public can participate, mass media efforts by themselves generally don't yield results. Upton Sinclair wrote *The Jungle*, which was an exposé of the meat-packing industry, as well as other food processing companies, a long time ago, but the workers at meat packing plants still face many of the same dangers that he publicized. The organizational structure for public involvement on that particular issue simply never existed.

Another favorite vehicle for well-meaning effort is the voluminous technical report and its variations, the lengthy technical testimony, the endless learned lecture, and all manner of impressive charts, data tables, credentials, degrees, statistical tests and probabilities, graphs, chemical formulae, sampling results, professional opinions, and polysyllabic verbal or written discourse. Again, these studies and technical reviews and analyses are necessary but insufficient.

Many academicians and citizens think that if a study is published and sent to government officials, then the truth in that study will triumph and everything will be set right. Some of these well-meaning study-senders will actually call a legislator or regulator and ask, "Didn't you read my study? Why isn't something being done?" Typically the answer will be, "Well, the plastics industry did their own study and their results were different from yours. I can't do anything if there are arguments among experts."

If there is no organization with clout supporting the results of the study, chances are high that it will be forgotten. There are roles for studies: those submitted by strong, active organizations get read and considered. These strongly backed studies have a chance of leading to changes.

Today, an organization must have plenty of political strength to institute change. To make a lasting change, information on the subject must be spread through both the local and national culture. We must take the trouble required to pursue lasting changes to guarantee safety and health and to safeguard the environment.

This book has emphasized the physico-chemical aspect of the plastics problem. Because culture and its role in political processes is under discussion, let us make a small digression into the other aspect of the plastics problem—namely the sociopsychological, anthropological aspect. Social scientists look at human cultures as whole entities and investigate the material as well as the social aspects of culture. Certain materials are characteristic of certain societies for their buildings, their clothing, their modes of storage, their decoration, and their religious artifacts. This is partly because of availability and partly because of historical forces or tradition.

Western culture is not the first society to use a disguised substitute material (plastic) for natural materials. The peoples of the Andes and Amazonia sculpted exquisite ceramic conch shells that sounded like real conch trumpets. Archaeological remains of some of these instruments are over a thousand years old. The ceramic used by these people was not inferior to the natural shell as a material for an instrument, and the artist's skill in making it resemble the real shell demonstrated that this was a high culture with high standards and techniques in craftsmanship.

The substitution of plastic for natural materials and real solids is having the *opposite* effect on our culture and our techniques of craftsmanship. Our workers are losing the techniques required to work with the real materials as they adjust to working with plastics. Plumbers with a full knowledge of welding techniques are becoming rare in certain areas. Because plastics are so easy to work with, unskilled, unlicensed plumbers and other construction and maintenance workers proliferate and often cause major damage.

Because of the oneness of culture, we may need to reflect on the social, psychological, and even spiritual implications of the invasion of this material. Plastic seems to be many things, but isn't, and it is inherently inferior to the natural materials and real solids being driven from the market. Is it possible that all this political hype, these cosmetic non-solutions to real problems, and the quick changes in veneer that are performed on cue by our national leaders are reflected by and reflect in our material culture?

If we find ourselves in a society that is plastic materially, socially, and politically, then the effort to reduce inappropriate and dangerous uses of plastics must be conducted very carefully. This effort must not encourage or make use of "plastic" behavior or political techniques. For example, simply pursuing a mass media campaign and giving interview after interview without doing any real organizing would be pure plastics.

What follows is an analysis of solid political and community programs, the political and social equivalents of welding steel and weaving wool rugs. These programs can leave an enduring legislative and regulatory legacy. Like welding steel and weaving wool rugs, these programs require patience, skill, training, and experience. They are not instant hypes.

## PROGRAMS FOR DRAGONSLAYERS

If we haven't got the big bucks of the plastics industry and we want to limit uses of plastics to appropriate safe applications, we have to attract human resources, individuals, and organizations. Even before we make any decisions about the laws and regulations we want, we have to create a network of people drawn from a broad spectrum of backgrounds.

One of the first things to do is to write out as long a list as possible of individuals and organizations, occupations and recreations, interest groups and advocates—all of whom may have some possible reason for increasing the safety of buildings by limiting the dangerous uses of plastics. The obvious people must be included on this list (firefighters, senior citizens, construction workers, public health advocates), as must the not-so-obvious people (theologians with environmentally oriented ethical codes, people who live around plastics factories and can be organized against the environmental problems therefrom, injured survivors of plastic fires, and environmental organizations).

We should brainstorm when writing this list. The more names of individuals and organizations it includes, the better. Don't forget hospital workers, hotel and restaurant workers, civil servants in "plasticky" skyscraper government buildings, parents of school children who come home ill because of fumes from plastics decomposition in a new school, mass transit riders, block associations in areas of much "plasticky" house rehabilitation or renovation, and building officials.

The list should eventually become a looseleaf notebook, because it will grow and grow. As the organizing moves forward, the people on the list who become involved will add their own lists of contracts. The list must eventually be computerized. The construction workers will bring in the contractors who don't want to use plastics. The injured survivors will bring in their doctors and lawyers. The environmental groups will bring in professors from campuses on which they have chapters. Eventually, the list may occupy more than one floppy disk, and in hard copy, two notebooks.

Soon after the list is initiated, a core group of trustworthy, faithful individuals must be identified. A steering committee should be formed. Candidates for this core group should have certain characteristics, such as a long record of fighting dangerous uses of plastics, and past contributions of money or political effort in this area. They should also represent a group directly affected by the dangerous uses of plastics or a group that acts as advocate for those who are directly affected, who have risked job or future financial security in their effort, and who make remarks about being ready for a long, tough haul.

This core group must include labor advocates for the environment, victims, and professionals including lawyers, scientists, and building engineers. The fundamental goals for legislation and regulation (as well as the strategy) will initially come out of the core group. Although some of the members of the core group will leave and new persons will join, there must be a certain level of stability for the people to work together and to rely on each other, and to maintain continuity of goals and strategy. This stability is another reason for choosing core group members who have a several-year record and who see the process as a long, tough haul. These are the people who will be around for years.

Besides the core group/steering committee, the organization needs one chairperson, at most two co-chairpersons. Someone must be responsible for getting things done and for communication, and must be a contact for the organization. The chairperson must be willing to do all kinds of work, and lots of it. He or she must be highly motivated and energetic.

The chairperson will probably be the one originating the idea of the organization and calling the first meetings. How can these first meetings come into existence? Begin by taking out the lists of "probably interested" individuals and groups. Pick up the telephone and talk with these people about plastics and the health and safety risks posed by plastics. Ask each one about his particular interest in the problem. Write down the problems they have had with plastics and the potential problems they foresee. Tell them you plan to call a meeting to organize on the issue and ask if they want a meeting notice. Also, ask them for the names and phone numbers of other individuals and groups who may have an interest.

There is no substitute for conversations and exchanges. The initial organizing involves making so many telephone calls that you won't welcome the arrival of the phone bill. Organizing locally can run up a large phone bill, but when compared with organizing regionally and nationally, it's not that large. This is similar to what the signers of the Declaration of Independence meant when they pledged "our lives, our fortune, and our sacred honor." We say, "time, money, and reputation."

After you have made all of these initial phone calls, you will have made a number of gains.

- You will know several individuals and organizations with potential for being active and working together.
- You will have learned more about the plastics issue as different people tell you about their experiences with plastics and with the politics and economics of the issue.
- You will see how the various aspects of the issue fit together.
- You will be known by a certain circle of people who will give your name and number to others; from your numerous phone calls, you will come into contact with especially interested people acquainted with the individuals and groups you called. You are almost ready to call a meeting, but not quite.

Before you call a meeting, you must visit face-to-face with selected people on your list. Some people and groups are so important that you must woo them and get their backing very early. Find out how they think the first meeting should be run and what its agenda should be. So brush off your good suit and polish your shoes—it's dress-up visit time.

The groups and people selected for visits differ from locale to locale. In nearly every city, the firefighters' and the pipetrades' unions merit a visit. These unions have national positions on plastics, toxins, and occupational and public health. They have experience and knowledge. The local chapters of environmental organizations often have members with knowledge, experience, and interest in issues of toxic exposure, who can offer a broad view of the spectrum of impacts resulting from a given substance or class of substances. In most areas there are victims of exposure to plastics fumes. These include workers in factories, people who have inhaled smoke during a fire, incinerator workers, and people who live around a polluting factory or around a plastics factory dump. Each of these victims has a story that is very important and that should be told. The victims should have the opportunity to tell their stories and to fight back at the forces that made them victims.

Each locale has its own important community groups. In Niagara Falls, everyone working on environmental issues visits the Ecumenical Task Force, an interdenominational group with a religious focus on toxic and environmental issues. In Poughkeepsie, New York, visits are paid to the Clearwater and Scenic Hudson organizations. Most of these local groups have run headlong into some form of the plastics issue and have members with interest, experience, and knowledge. Maybe someone is burning garbage to avoid garbage disposal fees and is making a neighborhood sick. Maybe people had to be evacuated when a warehouse that was storing plastics caught fire. Whatever the event, the local group will have someone who researched the details and has a file of articles and interviews.

When the calls have been made, the letters mailed, and the visits finished, and after the core group has formed, call the first meeting. Many of the interested organizations have meeting rooms and would be happy to host the meeting, if the date is set far enough ahead that the room can be easily reserved. The meeting notice should be sent out two to three weeks before the meeting, and the people who seemed especially interested and active should also be reminded about four to six days before the meeting. Three-quarters of organizing is phoning and mailing.

First meetings on important issues usually generate a decent response. People from widely different groups and backgrounds will attend a first meeting on the issue of dangerous uses of plastics. Many of these groups may have supported opposing sides of other issues, and don't exactly love each other. At the first meeting, it is critical to strike a balance between getting business done and letting the people get acquainted. One good way to begin the first meeting is to have everyone present introduce himself/herself and explain his or her particular interest in the issue. Give people time to tell their stories and to respond to the stories and statements of others. This serves a number of functions: it fosters cohesion, broadens the knowledge and outlook of the participants, and underscores the common issue and deflates the other possibly divisive issues. It also lifts the participants' morale to learn that all these other people have been concerned and are working on the same issue.

At the outset of the meeting, someone should be appointed secretary and take the minutes. From the introductory statements and from the core group's analysis of the situation, some parts of the program should be proposed at the very first meeting. The minutes will reflect the relationship between the introductory statements and the early proposals for a program.

The meeting chairperson has several tasks, all of which are vitally important in getting the group started functionally. These tasks are to welcome everyone and make everyone feel important, to encourage networking within the group, to make the group generate a few program proposals, to set up plans for the next meeting (time, place, and agenda), and to assign any necessary work. The members should get used to doing work for the group. Funds will have to be secured for mailings, copying, and other necessary services. The chairperson must bring cookies and apple juice to the first meeting. Do not underestimate the importance of refreshments!

After the official meeting, if all went well, the participants will stay and socialize. The chairperson can gauge how well the meeting went—if people hang around so long that the next group to use the room has to evict your group, then it went well.

Subsequent meetings should be regular. The program will develop and goals will be formed that will lead to significant improvements in safety and health. The means for furthering these goals will also grow out of the meetings and the between-meeting conversations and work. Because of the broad base of the group and because of the composition of the core group, sane and progressive goals and intelligent means toward them are almost automatic, although they will only be reached by hard work. Eventually, the group will arrive at a consensus on a program.

A typical program has long-term goals, legislative and regulatory means to those goals, public education and political activity to get the legislation and regulation, and research and writing to support both the drafting of the laws and statutes and the public education/politics. A good program sparks the supporting groups in several ways:

- It activates members who had been passive.
- It teaches new skills to members, broadening the outlook of the group.
- It shows how the particular concerns of the group fit in with those of a wider spectrum of society.
- It reinforces the public interest sense of mission that all nonprofit groups must have.

Because of the balance of influences of the core individuals and the larger organization, the program that brings a consensus will be good, if imperfect. The imperfections will become evident and, if the organization has gotten into a working relationship, will receive attention and correction.

A good program will also help attract contributions. Money is a continual problem for a loose coalition. There are standard ways to handle shaky finances. These include using the phones and mailing privileges of the supporting groups, getting foundation grants, finding individual wealthy "angels," and wheedling grants from the industries that rival plastics. Rival industries can do a lot, including providing hard-to-find information, planting news stories, and giving the use of their large public relations capabilities. However, accepting this help can leave the organization open to charges of being a front for these rival industries. Although this accusation has little power, it may cause some discomfort in inexperienced individuals. Because of the educational and political activities of the various industries, any organization that undertakes a program that addresses the dangerous uses of plastics will eventually receive offers of help from rival industries and offers from the plastics industry to "correct misinformation." Policy should already be in place about how to respond to these overtures.

The plastics industry will also probably insert a "ringer" into the organization. The identity of this person or persons will become obvious. They will give themselves away because of the particular courses of action they suggest. Even the core group may become infiltrated. In an organization formed with the purpose of curbing the dangerous uses of plastics, these people stand out like sore thumbs. The experienced veterans of previous plastics wars will be the first to smell the rats. Knowing looks will pass between these veterans when the ringers make certain telling statements that are straight out of the plastics industry's public relations campaigns. Eventually, everyone will know who this ringer is, and the only role the ringers can play is to keep the plastics industry informed of the organization's plans, because no one will listen to the suggested courses of action from the ringers.

Finally, the organization is ready for its legislative and regulatory campaigns. These grassroots campaigns must educate and activate the public and the government officials. To educate the public, your group can give press conferences, media events, and teach-ins, set up information tables in shopping areas, and send speakers to community organizations (PTA, churches and synagogues, senior citizens groups, block associations, tenant organizations, union meetings, etc.). To enlighten government officials, people from the organization who span the whole spectrum of its constituents should make lobbying visits. For example, the chairman of the State Senate Committee on Environment might see the representative of the plumbers' union sitting next to the representative of the Sierra Club. Perhaps they're both nodding in support of statements made by a mother who lost a child to toxic smoke inhalation. What the legislator sees is votes and constituency. If the Health Commissioner gets letters on many different letterheads, from the transit workers' union to Greenpeace to the Interfaith Committee on Toxic Hazards, that commissioner is under real pressure, and whoever appointed that commissioner feels that pressure.

The plastics industry will fight against any possible legislation with money, threats, and a media campaign. This industry will try to smear the leaders of the organization, threaten governmental officials with lawsuits, and entice elected officials with election funds and regulators with promises of jobs.

The correct response, of course, is to continue pushing to curb dangerous uses of plastics. Keep getting information to the public, keep holding letter-writing parties, and keep visiting legislators and regulators in motley groups that show broad constituency. Solid persistence is half the battle. Just keep showing up in mailrooms, offices, meetings, and conferences.

Some legislators and regulators will support the program and begin making law and statute. Give these people audiences and media coverage. Hold strategy sessions with their aides. A bill or statute may take years to be passed or approved, but those with persistent and broad support usually get on the books. The friendly legislators and regulators can give crucial advice to shorten the time and to win the end game.

When it becomes obvious that the public wants the law or regulation and will keep pressuring for it without becoming bored and simply giving up, then action is taken. This is what you are working toward—the end game. If the end game is not played carefully, the law or regulation will be purely cosmetic and not address the problem it ostensibly solves. The old FTC urethane foam settlement with the plastics industry (see Chapter 9) is a prime example of an action that made no one safer and did not regulate dangerous uses of plastics. The end game involves weighing proposed compromises in the wording of the law, in the classes of plastic to be covered, the types of occupancy to be protected, how it will be enforced, and the penalties for transgression. Some compromises will ease costs for building owners without seriously diluting the effect of the legislation or regulation and some compromises will appear innocuous but really gut the entire package. You must be shrewd and perceptive. The end game is trickier than any of the activities that led up to it. The combined skills and experience of the organization and its governmental friends must be mobilized to assure a real step forward in curbing dangerous uses of plastics.

## HOW WE GOT COMBUSTION TOXICITY INTO THE NEW YORK STATE BUILDING CODE

The New York State Building and Fire Prevention Code is at present one of the few codes that includes consideration of the differences in toxicity of different materials in a fire. The code includes a bank of combustion toxicity data based on the results of testing materials with the University of Pittsburgh protocol. Materials used in building systems and in finishings (floor and wall coverings) must first be tested and the test results reported to the State Office of Fire Prevention and Control. The data bank is available for builders, architects, building managers, and others who make decisions on materials to be used in buildings. These decision-makers are expected to consult the combustion toxicology data bank. Failure to do this could result in potential liability, if poor choice of materials leads to avoidable injuries and deaths in a fire. These decision-makers cannot claim ignorance anymore; New York State has the data to help in the choice of materials.

The fight to place the combustion toxicity data bank in the State Code lasted four years, 1982 through 1986. The plastics industry, the real estate industry, and many manufacturers of products fought it all the way, and succeeded in weakening the original concept. Originally, nearly all materials used in buildings would have been included in the Code—furnishings as well as building systems and finishings.

The idea for such a data bank started in January 1982. My husband, Rod, was waiting for a Broadway local subway at the Lincoln Center station in New York City. He saw a strange gray pipe and went over to inspect it closely. Written on the pipe was: PVC, 90 C. This pipe was electrical conduit for the new speakers in the public address system. Rod was with a friend and told him, "I'm going to expose this. This is outrageous!"

A friend of a friend was a photographer for the *New York Times* and went down into the stations to see the PVC conduit. He got the mass transit reporter, who at that time was Ari Goldman, interested. Goldman began investigating and found out about PVC and its fire behavior. His investigation resulted in a front-page photograph of Rod's face, grimacing at the conduit in the Lincoln Center station, in the February 2, 1982 issue of the *Times*.

The story triggered labor and political activity. The fire service unions, plumbers' union, and transport workers' union held press conferences and applied strong pressure on the politicians. The local community boards passed resolutions against the conduit. Straphangers, Inc., an organization formed by Ralph Nader, organized other groups, and they picketed the Transit Authority headquarters. I was asked to speak before the Metropolitan Transit Authority's Citizen Advisory Board about the technical details of fires in PVC. Finally, under the leadership of City Council President Carol Bellamy, the City Council found $2 million to replace the PVC with proper steel conduit in underground stations.

The conduit would have needed replacing anyway, because the heat that accumulated under the station ceilings from the trains and lights made it creep (sag and twist). (See Figure 1.1.) Because of the publicity and the growing flap about PVC in the subways, several organizations contacted us. We received a small grant from a steel corporation to be used to educate the public in New York State about what plastics do in fires, and we were called by several labor and community organizations.

Because of the tragedy at the Westchester Stouffer's Inn, the first statewide non-voluntary building and fire prevention code was being drafted. This tragic fire sensitized the public to the fact that plastics generate toxic fumes and smoke in fires. The time was right for organizing around the building and fire prevention code.

We began meeting with several labor unions. Together, we were able to organize quickly enough to be prepared for an important hearing that was being held. The purpose of the hearing was to discuss a bill that would fund a study on the feasibility of dealing with combustion toxicity in the Code. The testimony, follow-up letters, and lobbying created a network on which we could build.

We began visiting environmental groups in New York City, like the Natural Resources Defense Council, Environmental Defense Fund, Sierra Club, and the OSHA-Environment Network. It was a good time to visit these groups because the whole issue of the dangerous use and disposal of hazardous materials was coming to a boil. We left information packages and received information packages.

After the friendly visits, we started calling meetings. We continued to phone and visit other unions and environmental groups. The New York Lung Association's staff representative to the OSHA-Environmental Network became active in our group and helped with the meetings. Straphangers' Inc. was helpful. The Environmental Action Coalition gave us lists and lists of organizations and individuals.

The group decided to hold a statewide conference in Albany in October 1982. To generate representation from other areas of the state, we took our public education grant and used it to hold meetings and press conferences in the major cities of Albany, Buffalo, Rochester, White Plains, Binghamton, Syracuse, and Poughkeepsie. The organizations based in New York City had contacts in all of these other cities. Meetings were held in offices of labor unions and environmental groups.

During this public education/media blitz through New York, we met with hard working groups. These meetings broadened the perspective of the coalition. The groups we met with are given below as an example of the kinds of organizations you can expect to work with.

- The Love Canal Homeowners' Association and the Ecumenical Task Force of Niagara Falls/Frontier, which educated us all about the consequences of manufacturing wastes and about the spiritual and ethical side of the issues.
- The firefighters' union in Rochester, which had struggled to find out why five cases of cancer appeared among the firefighters who responded to a fire in a landfill where the Eastman Kodak Company had dumped old film and old chemicals.
- The Sierra Club locals all over the state and the state headquarters in Albany, which had received calls from all over about water pollution, air pollution, abandoned drums, and draft discharge permits.
- The pipetrades' union throughout New York, who had sponsored experimental burns of plastic plumbing beginning in the late 1960s and taught us about buildings and pipe.
- The building officials who desperately needed the means to prevent another fire like those in the MGM Grand Hotel and Stouffer's Inn.

From the meetings, we saw that the conference would have a broad agenda, with speakers addressing the plastics issue from many perspectives. These perspectives included those of the victims of fires and of toxic dumps, occupational hazards, legal issues of fires and of toxic dumps, the ethics of the issue, and the technical explanations of the risks.

After the first two cities were visited, we began to put out a newsletter with information gathered from the visits. As we made more visits and as more groups became involved both in New York City and statewide, the newsletter filled with information and events from all the perspectives represented in the coalition. As the different groups became bonded to each other through the newsletter and the meetings, they began to call each other for help. Even small typed newsletters without a professional look can knit a coalition together.

Planning and publicizing the conference took about two months. Our speakers came from as far away as the Washington, D.C. area. The program included experts, victims, and representatives of vulnerable groups such as transit workers. About a hundred people attended. Approximately ten were from the plastics industry.

The energy generated at the conference arose partly from the speakers, partly from the interaction between attendees, and partly from the pronouncements of the representatives of the plastics industry. These gentlemen parroted the evasive, misleading half-truths that were described in the previous chapter, again trying to mislead the public and government officials. But because the speakers had educated the participants, and because some of the participants had direct experience with plastics and spoke about it from the floor, the insidious nature of the standard lines of the plastics industry triggered strong reactions from the participants.

The last ninety minutes of the conference were devoted to obtaining consensus on legislative and regulatory policies. The conference decided on three issues.

1. To support the inclusion of combustion toxicity consid-

eration in the New York State Building and Fire Prevention Code.

2. To support the change in the statute of limitations for chemical-related personal injury suits.
3. To support the continued vacancy of certain land around Love Canal that was being considered for redevelopment into housing.

During these ninety minutes, the people from industry met in their own group, under the chairmanship of a senior vice president from a steel company. The group produced an industry statement on the problem of combustion toxicity. We established this separate group for industry mainly to allow the other groups to work without interference from the plastics industry. I assigned my husband, who has a Harvard accent, the task of inviting the industry representatives to produce a statement. He didn't believe me when I said that industrial executives are used to obeying authoritative voices. After he asked them to go into the northwest corner of the room and produce a statement, his eyes grew round with amazement as he watched them dutifully herd into the corner and sit down to work. He had just gotten obedience from vice presidents and division directors.

The industry statement turned out to be very interesting and even contained some admissions. When the plastics industry attempts to soften the code to allow replacement of metal or glass with plastics, we still use their statement, which says, "Organic materials by their very nature will burn."

After the conference, the newsletter served to coordinate our campaigns on the three consensus issues. Immediately, we generated letters to the Governor about the risks of allowing homes to be built and occupied in the ring around Love Canal. The network helped keep that land vacant for several years. Only now, in 1989, does it appear that the redevelopment boosters will get some of that land. By now, most people know about Love Canal and would factor that knowledge into any decision about buying a house on that land. We have to guard against misrepresentation of the risks by state and local government and the real estate industry.

In May of 1983, A.D. Little Corporation issued the report on the feasibility of addressing combustion toxicity in the State Building and Fire Prevention Code. The report recommended that the Code include a combustion toxicity data bank based on the University of Pittsburgh protocol. This inclusion, which would not actually regulate any material or any use, was offered as a first step in rational consideration of the combustion toxicity hazard.

The network, established in October 1982, worked for one year (May 1983 to May 1984) to pry the recommendation of the Secretary of State and place it before the State Code Council. The work included letter campaigns directed to the Secretary of State, letters to the Governor, and visits and letters to legislators. The plastics industry worked that year also, lobbying and threatening. The industry threatened lawsuits against the Code Council and also threatened the staff of the Office of Fire Prevention and Control (a division of the Secretary of State's office) with loss of their jobs.

Allies on the Code Council worked with us. The Executive of Dutchess County introduced a resolution passed by the Council that set aside a section of the Code for combustion toxicity considerations. The Code was printed up with a page indicating that this section would be inserted at a later date. The representative of the building officials on the Council brought up the section at each meeting.

Finally, the Secretary of State introduced the recommendation and resolution. Immediately, the plastics industry began raising objections. These objections appear in Chapter 2. Our eyebrows were raised when the working staff of the Council, two men from the Department of Housing and Community Renewal, began to echo the objections straight from the industry's "fact sheets" and public statements. These two began to set up obstacles for the passage of the resolution, going out of their way to delay it, and gave every sign that they received some benefit in working for the plastics industry. They even brought in someone from the National Bureau of Standards to lobby for the NBS combustion toxicity test method. This occurred shortly after a conference that exposed the inadequacy of the NBS test method. This inadequacy was revealed by the conversation between the NBS toxicologist and the inventor of the Pittsburgh protocol at an open scientific conference.

These two delayed the inclusion in the Code by bringing up the possibility that a regulatory impact assessment might be needed. The regulatory impact assessment took several weeks. Then public hearings were held, and each side had to marshall all its forces. The groups were faithful, and unions and environmental groups gave testimony all over the state.

When it came time for them to speak, the various industries cried poor. The cost of sending experts to testify at the hearings probably exceeded the cost of testing a product. But the crying did no good. The hearings ended with obvious widespread support for the insertion into the Code. The key here is widespread support.

In 1986, the Code Council met with great ceremony in a large auditorium that was filled with labor leaders, environmental groups, public health group officers, and building officials. Before the TV cameras and under the hot lights, the Council voted unanimously to insert the combustion toxicity data bank into the New York State Building and Fire Prevention Code.

Because of the widespread support, the Council members orated at length about the historic event. The Council chairman carefully assured that any Council member moved to speak had the opportunity. Of course, the audience applauded loudly.

I had gone to Albany earlier that day with plumbers and building officials from Westchester County. After the vote, they took me to a "lobbyist" restaurant (with very good food and wine), where we met plumbers and building officials from the rest of the state. A solemn toast was drunk to public health and safety. It had been a long battle, but the comrades stayed with it. Everyone had war stories, and some had battle scars. When the plastics industry uses its public relations apparatus to smear opponents, it can cause humiliation, threat to income, actual loss of income, and loss of credibility. The public relations experts have ways of presenting facts which imply terrible, untrue things about a person. This slickness can take a toll.

While the data bank war was fought, the other issue in the New York State toxics revolution also saw action. This issue was the change in the statute of limitations for personal injuries due to chemical exposure. Every year for six years, labor unions, victims, and environmental groups lobbied and organized for passage of this legislation. This activity culminated on Lobby Day in the spring, around the time that the committees were voting on this act. For the last three years of lobbying, extensive publicity turned the public against the Republicans who were controlling the State Senate and who kept the bill bottled up in committee. The Coalition for Toxic Victims' Justice, founded by labor unions and NYPIRG (New York Public Interest Research Group), held press conferences around the State. Present at these press conferences were victims who told stories about how their injuries did not become apparent until years after they were exposed to toxic chemicals, and how they could not get justice in New York State courts because of the ridiculous statute of limitations.

During the last three years of the fight, this Coalition broadened to include those people who had attended the October 1982 conference—firefighters, plumbers, other construction workers, toxic dump victims, etc. When uniformed firefighters along with uniformed nurses, began showing up on Lobby Day in busloads and going from office to office in the senate wing of the Legislative Office Building, even the Republican senators knew that they would have to make a deal soon. Finally, in 1986, the bill passed. It was signed into law in July of that year.

## WHAT IT TAKES TO GET RID OF LETHAL DRAGONS

These days, St. George wears many unlikely disguises, construction workers' overalls, Sierra Club hiking boots, the stethoscope of the occupational physician, and the high-heeled pumps of a secretary who works in a building furnished and finished with plastics. Just as the dragon lurks almost everywhere people live, work, or have fun, potential St. Georges are everywhere.

Today, it takes more time, effort, and money than ever before to get the legislation and regulation necessary to save ourselves from the dangerous uses of plastics. And the economic, social, and political factors behind this need multiply all the time. Only a major multifaceted reform would turn around this situation. However, even within this context, it is possible to make progress in these well-defined issues with persistence, hard work, commitment, and fellowship. Progress also requires the ability to endure the corruption and incompetence in our government without becoming bitter or enraged.

If the dragons are to be routed, we must consider the safety and health of our fellows and ourselves to be more important than short-term savings, the desire for objects, and convenience. What is your life worth? How dearly do you value your health? How important are the lives and health of your family, your friends, and all the people who inhabit your world? If enough people refused to buy dangerous forms of plastics and banded together to refuse to work with plastics under dangerous circumstances, many dragons would simply vanish.

For the types of plastics that have driven other materials out of the market, legislation and regulation must be passed that will restore public health and safety. Polyvinyl chloride wire insulation and urethane foam furniture stuffing and mattresses are prime examples of pervasive materials that are essentially without present alternatives. This is because manufacturers have not had to pay the true costs of using these products. Lives have been lost because of this. The safer materials can be resurrected if proper regulation is passed and if the price of these materials would actually reflect the true cost of the material to society.

This chapter outlines the activities necessary for passing the needed legislation and regulation. Just a small bit of regulation such as the combustion toxicity data bank, which does not really regulate the use of any material, takes years of effort and contributions. These are not activities for people with romantic ideas about overnight success. These are not activities for people who are sensitive to insults and smears. These are activities for people who place high value on lives and health and who maintain an enduring commitment to the well-being of people and a safe environment.

The issue of plastics, especially dangerous uses of plastics, confronts us all. There are contradictions between the way we live, the way we interact with others, and our professed value system. We must stop and think just how much value we actually place on life. We buy many unnecessary products that generate toxic waste and toxic air and water discharges at the factory. What about products that raise the probability that life and health will be lost during their use and disposal? Can we honestly say we love our children if, collectively, we leave them a world threatened with environmental catastrophes, like acid rain, loss of the ozone layer, the greenhouse effect, and climate change? What will we do about frequent mass disasters, including fires in large quantities of plastics? Can we even say we are concerned about the economy if we allow the disappearance of major industries that manufacture and process the traditional safe materials?

On the other balance of the scale are the actions that can be taken to resolve our crisis. These actions gleam like gold. The labor leader who risks his position by telling his constituents that he'll be negotiating as hard for a safer workplace as for higher pay and more holidays takes a gleaming action. The mother who pulls her children out of a school where decomposing plastics are making her children sick, and keeps them out in spite of threats by authorities, takes a gleaming action. The building official who refuses to soften the local code and allow plastic plumbing in spite of threats and pressures from the mayor's office takes a gleaming action. All over this country, the most unlikely St. Georges are wielding the most unlikely swords—voices, pens, feet, telephones, and cookies. St. George labors from California, where the people are fighting against plastic drinking water pipes, to New York, where the battle over the building code is fought continually.

The plastics industry pours money into these controversies and fights against all attempts to ban the dangerous uses of plastics. That is the advantage of economic capital. There are other forms of capital that also have their influences—social and political capital and the accumulation of human interactions on individual and group levels. Economic capital can be quantified. So many dollars are contributed to the legislator's campaign fund. Social and political capital is harder to quantify, although sociologists and political scientists can crudely measure human bonding. For example, when mixed groups drawn from a wide range of organizations visit a legislator or regulator, that government official has the fine problem of evaluating their bond by estimating the social and political capital accumulated by this coalition. He looks for clues like private jokes, nicknames, and the smoothness with which the individuals work together during the visit. If he sees cohesion, commitment, respect, and unity among the group, he knows that there is power to be exercised and the will to exercise it. Like economic capital, social and political capital are sources of power.

Accumulating social and political capital, like the accumulation of economic capital, takes hard work, drive, intelligence, and the willingness to risk resources. This undertaking is a true vocation, a calling that does not come to everyone. Not everyone can or will give up leisure time and spend hours on the telephone or at meetings. Not everyone will put cash into a fight for health and safety. But a surprising number of people do get the call and answer, like the prophets, "Here I am." They understand that these dangers affect everyone and that no one is immune to the dangers of smoke poisoning. Like the prophets, these activists are visited with apocalyptic visions, visions of likely scenarios of vast destruction and death if things remain the same and trends persist. Each of us wishes to protect our family and ourselves, but for these activists, "family" means *all* of Adam's children.

The dragon of myth embodies greed, death, and destruction. He sleeps on his treasure horde and makes lethal forays in his hunt for yet more. Many people have to be willing to fight him and even to die before he is finally killed. In order to conquer the dragons of dangerous plastics, we don't have to risk our lives. We're *saving* our lives. We have to be willing to buy materials that don't threaten our families and to work with people who are very different from ourselves and from each other. All of these people become our social-political family, our comrades. With comradeship and clear vision, we can effect change.

# Glossary

**Airpack**. Firefighter's term for the canister of compressed air on the self-contained breathing apparatus

**Aromatic hydrocarbons**. Organic molecules containing one ring-shaped structure or more. Examples include benzene, biphenyls, phenol, and toluene.

**Asphyxiants**. Chemicals that keep oxygen from either reaching or being used by body tissues.

**Autocatalysts**. Products of a chemical reaction that accelerate the reaction.

**Boundary Layer**. Layer of air or water (or any fluid) abutting on a solid and affected by friction with the solid. The fluid in the boundary layer flows more slowly than the main body of fluid and is often slightly chemically different.

**Carboxyhemoglobin**. Hemoglobin (the red oxygen-carrying protein in red blood cells) with carbon monoxide occupying its oxygen-carrying sites.

**Cathecholamines**. A class of neurotransmitter characterized by a benzene ring with a sidechain which includes an amino acid. Many automatic processes are controlled by cathecholamines, and many higher functions such as memory and mood.

**Dimethylformamide**. A powerful irritant present in the glue used by plumbers for plastic pipe.

**Feed Stock**. The chemicals that go into the manufacturing of materials and of other products such as pesticides.

**Fireball**. Word often used to describe rapidly moving, strongly directional fire such as the MGM fire.

**Fireproof**. Constructed of materials that do not burn, and built in such a way as to conform to certain standards. For example, a fireproof structure contains unit doors that will withstand a standard fire for either two or four hours, depending on the fireproof rating.

**Fire Retardants**. Chemicals that delay a material catching fire by raising its temperature of ignition. Fire retardants usually do not prevent heat-caused chemical decomposition.

**Halogen**. The elements fluorine, chlorine bromine, iodine, and the rare element astatine. Fluorine, chlorine, bromine are irritants.

**Irritants**. Chemicals that cause a sensation of burning, reddening, and inflammation. Corrosive irritants kill tissue and cause severe inflammatory reactions.

**Mastic**. A class of glue, often based on silicone.

**Natural Recent Polymers**. (As opposed to synthetic polymers made from fossil fuels.) Biologically synthesized repeating chains of molecules. Examples are wood, cotton, hair, fingernail, and fibrin. Natural polymers must be synthesized with a minimum of energy and are stable, low-energy materials.

**Neurotoxins**. Substances that are poisonous to nerves.

**Nitrocellulose**. Cellulose fibers treated with nitric acid or nitric and sulfuric acids so that the material then contains nitrogen. Nitrocellulose is very unstable in the presence of heat.

**Noncombustible**. Non-burning. Noncombustible materials are generally not organic, but are metal, glass, or cement.

**Organic Chemicals**. Chemicals primarily composed of carbon and hydrogen. They often contain oxygen, nitrogen, or sulfur also. They may contain many other elements as well.

**Plastic Resin**. The pure plastic as it comes from the reactor vessel before any additives are mixed into it. It is often a powder, shipped in bags.

**Plenum**. The space between the ceiling of one story and the floor of the story above it.

**Polymers**. Very long molecules composed of repeating units.

**Polymerized**. To synthesize a polymer out of its parent small units.

**Polymerization**. The process by which a large number of small molecules are joined together to form a large molecule, which is the polymer.

**Polyolefins**. A polymer (plastic) made up of small straight-chain units that contain only carbon and hydrogen.

**Polyethylene**. Polymer based on ethene as its parent unit. It is simply a very long chain of carbon atoms with two hydrogens attached to each carbon.

**Protocols**. Established methods for performing scientific projects; standard procedures for scientific tasks such as combustion toxicity testing, flammability testing, and smoke generation testing.

**Recently Grown Natural Polymers**. See Natural Recent Polymers.

**Resin**. See plastic Resin.

**Temperature of Ignition** (temperature of combustion). Temperature at which a material flames. It is slightly higher than that at which oxygen begins to combine with the carbon in the material and cause energy release.

**Temperature of Quantitative Decomposition**. Temperature at which a material begins rapid and predictable weight-loss due to breaking off and vaporization of side-chains and substitutions in the molecule. Lower than ignition.

# Bibliography

## Introduction.

*America Burning: Report of the President's Commission on Fire in America.* National Commission on Fire Prevention and Control. Washington, D.C. U.S. Government Printing Office, 1973.

## Chapter 1. Articles and books on the nature of plastics and the behavior of plastics in fires.

Aklonis, J.J. and W.J. McKnight. *Introduction to Polymer Viscoelasticity*, Second Edition. New York: Wiley, 1983.

Allara, D.L. "Aging of Polymers." *Environmental Health Perspectives*, 11, April 15, 1975.

Allara, D.L. and W.L. Hawkins (eds). *Stabilization and Degradation of Polymers.* Symposium Proceedings. Washington, D.C.: American Chemical Society, 1978.

Babrauskas, V. and J. Krasny. *Fire Behavior of Upholstered Furniture.* NBS Monograph 173. National Bureau of Standards.

Washington, D.C.: National Technical Information Service (NTIS), 1985.

Boettner, E.A., G. Ball, and B. Weiss. *Combustion Products From the Incineration of Plastics.* EPA Report #EPA-670/2-73-049. Washington D.C.: National Technical Information Service (NTIS), 1973.

Hall, J.E. and E.L. Tollefson. *A Literature Study of the Combustion Hazards of Polyvinylchloride (PVC) and Acrylonitrile Butadiene Styrene (ABS).* Report prepared on behalf of the Canadian Foundry Association, Cast Iron Soil Pipe Division, Edmonton, Alberta, Canada: 1981.

Hawkins, W.L. *Polymer Degradation and Stabilization.* New York: Springer Verlag, 1984.

Hilado, C. *Flammability Handbook for Polymers.* Westport: Technomic Press, 1982.

Kelen, T. *Polymer Degradation.* New York: Van Nostrand Reinhold Co., 1983.

Levin, B., M. Paabo, M. Fultz, and C. Bailey. "Generation of Hydrogen Cyanide From Flexible Urethane Foam Decomposed Under Different Combustion Conditions." *Fire and Materials*, 9:125–133, 1985.

Liao, J.C. and R.F. Browner. "Determination of Polynuclear Aromatic Hydrocarbons in Poly(vinyl chloride) Smoke Particles by High Pressure Liquid Chromatography and Gas Chromatography-Mass Spectrometry." *Anal. Chem.*, 50:1683, October 1978.

National Bureau of Standards. *Polymer Degradation Mechanisms.* NBS Circular 525, 1953.

O'Mara, M.M. "High-Temperature Pyrolysis of Polyvinyl Chloride: Gas Chromatographic-Mass Spectrometric Analysis of the Pyrolysis Products From PVC Resin and Plastisols." *Journal Polymer Science, Part A-1*, 8:1887, 1970.

O'Mara, M.M. "Combustion of PVC." *Pure and Applied Chemistry.* 49:649, 1977.

Paabo, M. and B. Levin. *A Review of the Literature on the Gaseous Products and Toxicity Generated from the Pyrolysis and Combustion of Rigid Polyurethane Foams.* National Bureau of Standards report #NBSIR 85-3224. NTIS #PB86-151941. 1985.

Troitskii, B.B., L.S. Troitskaya, V.N. Myakov, and A.F. Lepaev. "Mechanism of the Thermal Degradation of Poly(vinyl chloride)." *Journal of Polymer Science Symposium.* 42:1347, 1973.

**Chapter 2. Example references on fire spread.**

Blackshear, P.L. (ed.) *Heat Transfers in Fires: Thermophysics, Social Aspects, Economic Impact.* Washington, D.C.: Scripta Book Co., 1974.

Cooper, L.Y. "The Development of Hazardous Conditions in Enclosures With Growing Fires." *Combustion Science Technology.* 33:279–297, October 1983.

Denyes, W. and J. Quintiere. "Experimental and Analytical Studies of Floor Covering Flammability With a Model Corridor." *Floor and Floor Covering Materials, Fire and Flammability Series* Volume 12. Westport: Technomic Press, 1974.

**Example references on soot chemistry. The following references were cited in the Chapter 1 references.**

Boettner et al, 1973; Liao and Browner, 1978; and O'Mara, 1970 and 1977.

**Others are as follows.**

Iida, T., M. Nakanishi, and K. Goto. "Investigations on poly-(vinyl chloride) I. Evolution of Aromatics on Pyrolysis of Poly(vinyl chloride) and Its Mechanisms." *Journal Polymer Science, Polymer Chem. Ed.* 12:737, 1974.

Newman, R., A. Tewarson, and E. Farren. "The Effects of Fire-Exposed Electrical Wiring Systems on Escape Potential from Buildings. Part II— Comparative Fire Tests." *Factory Mutual Research for the National Electrical Manufacturers Association.*, 1976a.

Sarkos, C.P., J.C. Spurgeon, and E.B. Nicholas. "Laboratory Fire Testing of Cabin Materials Used in Commercial Aircraft." *Journal of Aircraft.* 19 (2):78, 1979.

Stone, J.P., R.N. Hazlett, J.E. Johnson, and H.W. Carhart. "The Transport of Hydrogen Chloride by Soot From Burning Polyvinyl Chloride." *Journal of Fire Flammability.* 4:42, 1973.

Tewarson, A. and R. Pion. "The Effects of Fire-Exposed Electrical Wiring Systems on Escape Potential from Buildings. Part III—Comparative Laboratory Tests." *Factory Mutual Research for the National Electrical Manufacturer's Association,* 1976b.

Alarie, Y. "The Toxicity of Smoke From Polymeric Materials During Thermal Decomposition." *Annual Review of Pharmacological Toxicology.* 25:325–347, 1985.

Alarie, Y. and B. Levin. Discussion on Levin's paper, "Conditions Conducive to the Generation of Hydrogen from Flexible Polyurethane Foam." *Proceedings of the Joint Panel Meeting of the UJNR Panel on Fire Research and Safety (7th),* 1983. U.S. National Bureau of Standards. NTIS #PB85-199545: 469–470.

Bukowski, R.W. and W.W. Jones. "The Development of a Method for Assessing Toxic Hazard." *Fire Journal,* March 1985.

Clarke, F.B. "Toxicity of Combustion Products: Current Knowledge." *Fire Journal,* 77:84, 1983.

A.D. Little, Inc. *Study to Assess the Feasibility of Incorporating Combustion Toxicity Requirements Into Building Materials and Furnishing Codes of New York State.* Final Report to Department of State, Office of Fire Prevention and Control. 1983.

National Research Council, Committee on Fire Toxicology. *Fire and Smoke: Understanding the Hazards.* Washington, D.C.: National Academy Press, 1986.

**Chapter 3. The New York Telephone Fire.**

Lee, V. and D. Singleton. "PVC Fire's Effect: Death?" *New York Daily News,* December 18, 1983.

Rodriguez, L. and D. Chapman. "The Night All the Bells Rang at Ma Bell's." WNYF, first quarter, 1975. (WNYF is the official publication of the New York Fire Department.)

Wallace, D., "Danger of Polyvinyl Chloride Wire Insulation Decomposition: I. Long Term Health Impairments: Studies of Survivors of the 1977 Beverly Hills Supper Club Fire." *Journal of Combustion Toxicology*, 8:205–232, 1981.

Wallace, D., N. Nelson, and T. Gates. "Polyvinyl Chloride Insulation Decomposition: II. Consideration of Long Term Health Effects from Chlorinated Hydrocarbons." *Journal of Combustion Toxicology*, 9:105–112, 1982.

**Reports on the fires in Chapters 4–8.**

The National Fire Protection Association issued reports on the following fires. These reports are available to the public for a fee.

Younkers Brothers' Department Store (1978)
MGM Grand Hotel (1980)
Stouffer's Inn (1980)
Ramada Inn (1983)

**Other available reports.**

The Des Moines Fire Department gathered all documents and reports on the Younkers Brothers' Department Store fire into two notebooks.

On the MGM Grand Hotel fire: The Clark County report on the investigation of the MGM Grand Hotel fire, the National Institute of Occupational Safety and Health's Health Hazard Evaluation for the firefighters, and the National Bureau of Standards' Fire Research Center's report on the analysis of the blood gases and of the soot elements.

On the Stouffer's Inn Fire: The Report of the Governor's Panel to Investigate the Stouffer's Inn Fire, also called the Blue-Ribbon Panel's Report.

On the PATH fire: The report of the Port Authority Trans-Hudson Corporation, *Special Investigation Report: PATH Car No. 725 Fire, March 16, 1982.* This report has an appendix also.

Also available is: National Transportation Safety Board. NYC-82-F-R032, "Electric Self-Propelled Passenger: Tunnel Fire and Evacuation." 1982.

**If you are interested in the connection between arson and sexuality, which is mentioned in Chapter 7, the following will be helpful.**

Scott, D. *The Psychology of Fire.* New York: Charles Scribner's Sons, 1974.

**For further reading, non-scientists will find the following articles easy to understand.**

Donovan, Lynn. "Polyvinyl Chloride: Fighting the Secret Killer in Fires." *RN Magazine,* February 1978, pages 59–63.

Mann, Charles. "The New Firefighters." *Science Digest.* 99:70–74, April 1984.

Michaels, Mark. "Tunnel Vision." *Firehouse.* 47–52, May 1986.

Schaeffer, Robert. "Where There's Smoke." *Not Man Apart, The Newsmagazine of Friends of the Earth,* 12(9a):8–9, 1982.

Taylor, R.E. "Designing to Stop Smoke—The Real Enemy in a Fire." *Specifying Engineer.* 74–76, May 1982.

U.S. Consumer Product Safety Commission. *Fact Sheet No. 41: Plastics.* 1975, GPO 889–731. Three-page pamphlet.

# Index